The
Catharanthus
Alkaloids

The Catharanthus Alkaloids

Botany, Chemistry, Pharmacology, and Clinical Use

EDITORS

William I. Taylor

INTERNATIONAL FLAVORS AND FRAGRANCES
UNION BEACH, NEW JERSEY

AND

Norman R. Farnsworth

COLLEGE OF PHARMACY
UNIVERSITY OF ILLINOIS AT THE MEDICAL CENTER
CHICAGO, ILLINOIS

MARCEL DEKKER, INC. New York

MARCEL DEKKER, INC.

270 Madison Avenue, New York, New York 10016

LIBRARY OF CONGRESS CATALOG CARD NUMBER: 75-7710

ISBN: 0-8247-6276-2

Current printing (last digit):
10 9 8 7 6 5 4 3 2 1

PRINTED IN THE UNITED STATES OF AMERICA

CONTRIBUTORS

DONALD J. ABRAHAM, Department of Medicinal Chemistry, University of Pittsburgh, Pittsburgh, Pennsylvania

DAVID A. BLAKE,* Department of Pharmacology and Toxicology, School of Pharmacy, University of Maryland, Baltimore Maryland

DAVID P. CAREW, College of Pharmacy, University of Iowa, Iowa City, Iowa

WILLIAM A. CREASEY, Departments of Internal Medicine and Pharmacology, Yale University, School of Medicine, New Haven, Connecticut

R. C. DECONTI, Departments of Medicine and Pharmacology, Yale University, School of Medicine, New Haven, Connecticut

NORMAN R. FARNSWORTH, Department of Pharmacognosy and Pharmacology, College of Pharmacy, University of Illinois at the Medical Center, Chicago, Illinois

RONALD J. PARRY,** Department of Chemistry, Stanford University, Stanford, California

WILLIAM T. STEARN, Department of Botany, British Museum (Natural History), London, England

GORDON H. SVOBODA, Eli Lilly and Company, Indianapolis, Indiana

M. TIN-WA, Department of Pharmacognosy and Pharmacology, College of Pharmacy, University of Illinois at the Medical Center, Chicago, Illinois

*Present address: Departments of Pharmacology and Gynecology/ Obstetrics, School of Medicine, The Johns Hopkins University, Baltimore, Maryland

**Present address: Department of Chemistry, Brandeis University, Waltham, Massachusetts

PREFACE

The varied character of natural products, and indeed their
very existence, pose fundamental questions to scientists. While
many books have been published concerning the chemical aspects
of natural products, it is an exceptional case when most aspects
of a class of substances can be found within the covers of a
single volume. We have attempted to deal with the *Catharanthus*
alkaloids in the same manner as for the *Vinca* alkaloids, and
have brought together the work of experts in various specialized
fields of endeavor.

This volume was constructed with two groups of readers in
mind. The first is typified by graduates or advanced graduates
who are well grounded in organic chemistry, biochemistry, botany,
pharmacology and medicine. The second group is composed of
established researchers. Both groups can use this book either
for reference or as a text, especially for the convenience
afforded by this multidisciplinary approach.

This book was several years in preparation, due primarily
to several individuals who initially agreed to contribute, but
for a variety of reasons had to decline. To those contributors
who prepared chapters early, we are grateful for their patience
and understanding, and for being willing to prepare, in certain
instances, addenda to bring the material up-to-date.

Finally, this book could not have been prepared without
the dedicated typing efforts of Mrs. Anne-Grete Kreiborg (now
returned to Copenhagen), Ms. Pearl Spears, Ms. Janis Merkle and

v

Ms. Deborah Posley, to whom we owe a debt of thanks. The
structures, for the most part, were prepared by Dr. George H.
Aynilian, and Messrs. William Krutul and William Wheatcroft,
and we also would like to express our appreciation to them.

William I. Taylor
Norman R. Farnsworth

CONTENTS

The Catharanthus Alkaloids

INTRODUCTION

Gordon H. Svoboda

Eli Lilly and Company
Indianapolis, Indiana 46206

Botanicals have been a source of medicinal agents since
time immemorial. The history of herbal medicine in the treatment
of disease can be identified with the history of medicine and with
the history of civilization itself. It should therefore astonish
no one to find modern man delving into the folkloric past in hopes
of discovering new therapeutic agents to treat or cure human
diseases.

As cited in Chapter II, the folkloric pathway is fraught with
inherent pitfalls. Many diseases described in ancient literature
are for the most part unrecognizable in terms of modern diagnoses.
Furthermore, many folkloric remedies were used as panaceas,
involving conditions which cannot yet be duplicated in animal
systems.

Two laboratories, working independently of each other and
whose initial efforts were unknown to one another, had selected
the Madagascan periwinkle because of its reported folkloric use

1

as an oral hypoglycemic agent. Additional focus was also placed
on this plant in the author's phytochemical screening program
because other *Vinca* species were reported to contain hypotensive
alkaloids. (The plant under consideration, *Catharanthus roseus*
G. Don, was then more commonly referred to as *Vinca rosea* Linn.,
and no differentiation had been made between the genera
Catharanthus and *Vinca*.) Using various types of crude extrats,
neither group could, however, demonstrate experimental hypogly-
cemia in either normal or experimentally-induced hyperglycemic
rabbits.

In attempting to devise a new blood sugar assay, the
research group of Noble, Beer and Cutts at the University of
Western Ontario observed a peripheral granulocytopenia and bone
marrow depression in rats produced by certain select fractions.
Investigation of these active fractions eventually resulted in
the isolation of the dimeric indole-indoline alkaloid vincaleukoblastine (VLB) as the sulfate salt.

Submission of a defatted ethanolic extract of the whole
plant to our independent cancer screening program at Eli Lilly
and Company resulted in the observation of a profound and
reproducible prolongation of life for DBA/2 mice infected with
the P1534 leukemia, a transplanted lymphocytic leukemia. The
availability of this leukemia as a biological monitor enabled
us to follow purification steps which eventually allowed for the

initial isolation of leurosine, a dimeric alkaloid closely
related to VLB, as well as VLB.

The main focus, however, of the author's investigation was
on the isolation of the entity(ies), presumably alkaloidal, which
was responsible for producing laboratory cures, inasmuch as
neither VLB nor leurosine, nor any therapeutic combination thereof
was capable of achieving this result. This effort finally
culminated in the isolation of leurocristine and leurosidine,
two dimeric alkaloids capable of producing laboratory cures.

Classical extraction and purification techniques were of
little value throughout the course of our investigation. A new
technique of selective or differential extraction was devised.
This, coupled with another new technique, gradient pH, together
with a set of fortuitous circumstances, allowed for the isolation
of some 50 new alkaloids in our laboratories, along with three
others which were codiscovered in other laboratories. To date,
64 alkaloids, three of which we have never encountered, have
been reported as having been isolated from mature plants of
C. roseus. In addition, recent studies related to alkaloid
biosynthesis involving immature plants have resulted in the
isolation of five known monomeric alkaloids and two of their
derivatives, along with four glycosides. Therefore, a total of
75 distinct alkaloidal entities have been isolated from *C. roseus.*

Three distinct pharmacological activities have been seen
with a number of these alkaloids – hypoglycemic, diuretic and

antitumor. Six of the 23 reported dimeric alkaloids possess
experimental oncolytic activity, and two of these, vincaleukoblastine (VLB) and leurocristine (LC), have found extensive
application in the treatment of human neoplasms.) ✳

The volume of clinical literature is indeed most impressive –
to date more than 3000 citations have been recorded. The hoped-
for antileukemic activity of VLB has not materialized. It is
used fairly extensively in the treatment of Hodgkin's disease
and other lymphomas such as lymphosarcoma, reticulum-cell sarcoma
and advanced stages of mycosis fungoides, neuroblastoma, metho-
trexate-resistant choriocarcioma and Letterer-Siwe disease
(histiocytosis X); carcinomas of the breast unresponsive to
appropriate endocrine surgery and hormonal therapy and embryonal
carcinoma of the testis.

By contrast, the most striking clinical feature of leurocristine is its ability to induce complete hematologic remissions
of acute leukemias in children, both lymphocytic and myelogenous.
It has been referred to as a "miracle" drug – "Judged by the
usual yardstick of time for the development of new drugs and
their clinical acceptance by physicians, vincristine (leurocristine) qualifies as a miracle drug, for it was only 10 years ago
(1958) that work was begun on this compound. Except for the
increased activity in the field of cancer research, the work on
vincristine was not abetted like the work on the miracle drug
penicillin by a World War. In spite of this, vincristine

presently is held in high regard by cancer chemotherapists,
and much is known of its toxicology, pharmacology, and clinical
activity in the human being".

Other disseminated or non-resectable neoplasms which have
been reported as being responsive to leurocristine therapy
include lymphomas such as Hodgkin's disease, lymphosarcoma and
reticulum-cell sarcoma; carcinomas of the breast which are
unresponsive to appropriate endocrine surgery and hormonal
therapy, carcinomas of the cervix and of the prostate; choriocarcinoma, primary brain tumors (astrocytomas), neuroblastoma,
rhabdomyosarcoma and Wilms' tumor.

In contrast to other therapeutic agents leurocristine is
not myelosuppressive and is considered to be the ideal compound
for use in combination chemotherapy. A profusion of combination
regimens exist, most of which include leurocristine as a component,
and it is well documented clinically that such combinations induce
higher remission and curative rates than when using single agent
therapy.

The dimeric indole-indoline alkaloids represent a new class
of compounds unrelated to any other naturally occurring oncolytic agents. While the structures of most of the active antitumor
alkaloids have now been elucidated, much still remains to be done
to acquire a full understanding of structure-activity relationships. Structural modifications have yet to achieve an enhance-
ment of VLB activity with a subsequent reduction in leukopenic

effects; furthermore, a derivative of leurocristine has not been found which possesses reduced neuromuscular manifestations. Dimerization of naturally occurring monomeric alkaloids is being pursued in a number of laboratories and one can reasonably expect new compounds which will eventually be of clinical interest.

Leurosine is superior to both VLB and leurocristine in its capability in lysing a culture of malignant cells. However, its lytic ability is completely inhibited in the presence of adult human plasma and unaffected in the presence of fetal plasma. No such effect is seen with either VLB or leurocristine. Obviation of this protein inactivation in the human situation presents a significant clinical challenge. - Sufficient quantities of leurosidine have never been stockpiled to initiate and complete a full-scale clinical trial. The activity of leurosidine against the P1534 leukemia is superior to that of leurocristine and it would be interesting to ascertain whether or not its effectiveness would project into the human situation. Sufficient quantities, however, would have to come from an alkaloid modification program and said synthesis has not yet been achieved.

Data emanating from various laboratories studying the mechanism of action of the *Catharanthus* alkaloids appear to be conflicting both in terms of techniques and systems used. A concentrated effort at the molecular level, using drug-sensitive systems, appears to be warranted.

It has now been 11 years since the American Society of Pharmacognosy sponsored its Pittsburgh Symposium on the "Chemistry and Biological Activity of *Catharanthus, Vinca* and Related Indole Alkaloids." Recently a comprehensive review was published, one which deals with the botany, phytochemistry, chemistry and biological activities of *Vinca* species. This volume will be its counterpart for *Catharanthus* species.

Interest in both of these genera remains high and investigations are actively being pursued in a number of laboratories. This author anticipates the announcement of significant progress in several areas of research even prior to the publication of this volume. Perhaps we can look forward to another symposium sponsored by the American Society of Pharmacognosy, one similar to that of 1964.

CHAPTER I

A SYNOPSIS OF THE GENUS CATHARANTHUS (APOCYNACEAE)[*]

William T. Stearn

Department of Botany
British Museum (Natural History), London, England

The genus *Catharanthus* G. Don is a member of the botanical
family Apocynaceae and belongs to the subfamily Plumerioideae,
being placed in the tribe Plumerieae subtribe Alstoniiae of
K. Schumann's classification in Engler & Prantl's *Natürlichen
Pflanzenfamilien* IV Abt. 2 (1895) and the tribe Alstonieae subtribe Catharanthinae of Pichon's classification in *Mém. Mus.*

---

[*] This synopsis of the genus *Catharanthus* is dedicated to the
memory of Marcel Pichon (1921-1954) who published some 50
memoirs and notes on the Apocynaceae and was engaged at the
time of his sudden death on the preparation of an account of
the family for the *Flore de Madagascar*. The historical section here is an amplified version of a paper by the author
in *Lloydia* 29:196-200 (1966) and is used here by courtesy of
the Editor of *Lloydia*. The descriptions of the genus *Catharanthus* and *C. roseus* were originally prepared for Fawcett &
Rendle's *Flora of Jamaica* Vol. 6. A detailed account of all
the Madagascar species will be published by Friedrich Markgraf
in the *Flore de Madagascar* No. 169. Hence only the original
descriptions of the other Madagascar species are given below.

Nat. Hist. Nat. Paris N.S. 27:239 (1949). Its closest ally is
the north temperate genus *Vinca* L., in which Linnaeus and his
contemporaries and followers included the first known species,
C. roseus (Vinca rosea). Separated generically by Reichenbach
in 1828, by G. Don in 1835, and Endlicher in 1838, it was returned to *Vinca* by Alphonse de Candolle in 1844, kept there by
Bentham and Hooker in 1876, but restored to generic rank again
by K. Schumann in 1895. Most later authors have maintained it
as an independent genus, some using the generic name *Lochnera*,
one *Ammocallis*, others *Catharanthus*. It consists of seven
species (including *C. roseus*) endemic to Madagascar (Malagasy
Republic) and one, *C. pusillus*, to India. The best known
species, the Madagascar periwinkle, *C. roseus*, has long been
esteemed as an ornamental plant; it is easily cultivated, seeds
freely and in tropical countries becomes naturalized and spreads
rapidly as a garden escape; in this way, although originally
an endemic Madagascar species, it has acquired a pantropical
distribution. Recently it has become celebrated as the factory
of some seventy alkaloids, these including a number with demonstrable oncolytic activity and a few with actual clinical value
in the treatment of cancer. Thus it has stimulated much research
in pharmacology and biochemistry. Agreement on its correct botanical name is needed to ensure the recording of information
under one name instead of several.

     Since 1920 botanists dealing with Apocynaceae of the West
Indian flora have mostly adopted the name *Catharanthus roseus*,
which was used, for example, in Britton & Millspaugh's *Bahama
Flora* 336 (1920), by Britton & Wilson in *Scientific Survey of
Porto Rico* 6:88 (1925), by Hubbard & Rehder in *Botan. Leafl.
Harvard Univ.* 1:4 (1932) and in J. K. Small's *Manual of the
South-eastern Flora* 1060 (1933) long before Pichon took it up
in 1949. The name *Catharanthus* has also been used in such
standard works as R. A. Dyer *et al., Flora of Southern Africa*

26:267 (1963), Gooding *et al.*, *Flora of Barbados* 327 (1965),
Backer & Bakhuisen, *Flora of Java* 2:227 (1965), and Adams,
Flowering Plants of Jamaica 589 (1972).

In 1964, however, J. D. Dwyer concluded in *Lloydia* 27:282–
285 that "the legitimate name of the Madagascan periwinkle should
stand as *Lochnera rosea* (L.) Reichenbach." Thus he returned to
the nomenclature used, for example, by K. Schumann in Engler &
Prantl's *Die natürlichen Pflanzenfamilien* IV. 2:145 (1895) and
by Stapf in Thiselton-Dyer's *Flora of Tropical Africa* 4. i:118
(1902). An investigation of its taxonomy and nomenclature, undertaken in 1956 as part of the work on Fawcett & Rendle's *Flora
of Jamaica*, led me to accept the current view that *Vinca rosea*
and its close allies differ in so many characters from *Vinca*
proper that they should be put in a separate genus, that the
correct name for this genus under the *International Code of Botanical Nomenclature* is *Catharanthus* and not *Lochnera* and that
consequently the correct name for the Madagascar periwinkle is
Catharanthus roseus and not *Lochnera rosea*. A re-examination
of the facts has confirmed these conclusions.

VINCA AND CATHARANTHUS

The genus *Vinca* was established by Linnaeus in 1753 in his
Species Plantarum (1:209), where he distinguished two species,
V. minor and *V. major;* the generic description to be associated
with his specific diagnoses appeared in 1754 in the fifth edition
of his *Genera Plantarum*. *Vinca minor* L. is the accepted lectotype of the genus. These two original species, both European,
resemble one another fairly closely in floral structure. The
corolla is funnel-shaped with a gradually expanded tube; the
flattened filaments of the stamens bend sharply outwards from
the corolla-tube, then curve back to the tube so that the anthers with their apical appendages arch over the stigma; Linnaeus described the filaments as being "inflexa, retroflexa."
The flowers are borne on long axillary pedicels.

To this homogeneous north-temperate group Linnaeus added
in 1759 the tropical *Vinca rosea* (now the accepted lectotype of
Catharanthus) although it did not fit his generic description
as regards the stamens. His protologue in *Systema Naturae* 10th
ed., 2:944 (1759) reads as follows:

> *rosea.* ·B. V.   (=VINCA) caule frutescente erecto
> rigido, fol. (=foliis) ovatis. *Mill. fig.* 186.

This diagnosis, though brief, sufficed to distinguish *V. rosea*
specifically, but not generically, from the three other species
included by Linnaeus in *Vinca*, i.e. *V. minor*, *V. major* and *V.
lutea* (now treated as *Urechites lutea*). The reference by Linnaeus to Philip Miller's *Figures of the Most Beautiful, Useful
and Uncommon Plants* 2:124, t. 186 (1757) places its identity
beyond dispute, since Miller here gives both a detailed description and an excellent illustration under the name *Vinca foliis
oblongo-ovatis integerrimis, tubo floris longissimo, caule ramoso fruticoso.* This species is now pantropical and has even
been regarded as native to the West Indies. Miller's text set-
tles its origin: "The seeds of this plant were brought from
Madagascar to Paris, and sown in the King's Garden at Trianon,
where they succeeded; and from thence I was furnished with the
seeds, which succeeded in the Chelsea Garden." Introduced in
tropical gardens as an ornamental plant, it quickly became
naturalized. Thus Lunan's *Hortus Jamaicensis* 2:60 (1814) records both the red and the white forms as being cultivated in
Jamaica and adds: "They were introduced from the East Indies,
and thrive very well in Jamaica. The red kind, indeed, may be
found wild about the streets of Kingston and Spanish Town, and
in many other parts of the island." The apparently unfounded
reputation of this plant in Jamaican folk-medicine for the
treatment of diabetes led, however, to the discovery of its

numerous alkaloids, among them vincaleukablastine and leurocristine.

NOMENCLATURE OF CATHARANTHUS

The first to recognize *Vinca rosea* as being generically
distinct from *Vinca, sensu stricto,* was Ludwig Reichenbach, who
in 1828 proposed for it the generic name *Lochnera*. This separation is taxonomically sound. Pichon in 1949 listed 34 differences between *Vinca* and *Lochnera,* most of which are tabu-
lated and illustrated by Lawrence in *Baileya* 7:113-119 (1959).
Unlike *Vinca*, the latter has almost sessile clustered flowers
with the corolla salver-shaped, the tube is almost cylindric
and the stamens are almost sessile, with their filaments very
short and their anthers lacking terminal appendages. The ac-
companying drawing (Fig. 1) by Priscilla Fawcett illustrates
these floral differences. The seeds also differ, being 6-10
mm. long and brown in *Vinca*, and 1-3 mm. long and black in
Lochnera. Whereas *Vinca* has the chromosome numbers 2n=46 and
2n=92, all the species of *Lochnera* examined (incl. *Catharanthus
pusillus* and *C. roseus*) have 2n=16.

Unfortunately Reichenbach gave no characters whatever for
his new genus *Lochnera*. The relevant complete entry in his
Conspectus Regni Vegetabilis 1:134 (1828) is as follows:

> 3533.   Vinca. L.
> 3533 b. Lochnera. Rchb.
>   *V. rosea*. L.

This provides neither a description of the genus *Lochnera* nor
any statement of its difference from *Vinca*. *Lochnera* is here
a *nomen nudum* (a name unaccompanied by description); it is not
supported by any reference to a previously published description of the genus; it has no nomenclatural validity. As stated
in the *International Code of Botanical Nomenclature* 1961, Arti-

Catharanthus   Vinca

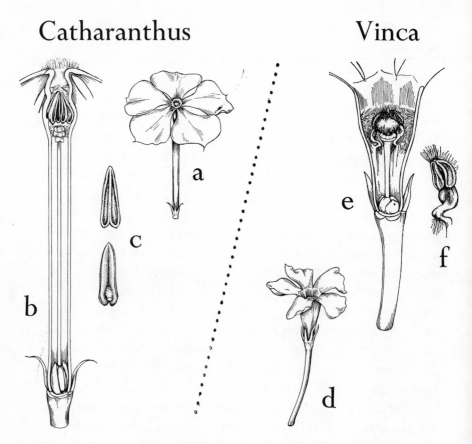

Figure 1.   Floral differences of *Catharanthus* and *Vinca*.
            Catharanthus roseus: *a*, flower, x 1; *b*, flower
            in section, x 3; *c*, stamen, x 6. *Vinca minor:*
            d, flower, x 1; *e*, flower in section, x 3; *f*,
            stamen, x 6.   (drawing by Priscilla Fawcett)

cle 34, "a name is not validly published . . . by the mere mention of the subordinate taxa included in the taxon concerned."
Moreover, as stated in Article 41, "in order to be validly published, a name of a genus must be accompanied (*i*) by a description of the genus, or (*ii*) by a reference (direct or indirect)

to a previously and effectively published description of the
genus in that rank or as a subdivision of a genus." The phrase
"subdivision of a genus" refers only to taxa above and not at
the specific level, i.e. between genus and species in rank (*Int.
Code* 1961, p. 26 footnote). By merely mentioning *Vinca rosea*
as a member of the undescribed genus *Lochnera* Reichenbach did
not indirectly refer to a description of that genus either as
a genus or as a subdivision of a genus. The main prop of Dwyer's
contention in favour of *Lochnera* against *Catharanthus* thus falls
to the ground. His others likewise collapse when touched. The
name *Lochnera* remained a *nomen nudum* until Endlicher provided a
description in his *Genera Plantarum* 583 no. 3406 (August 1838).
Here he correctly described *Vinca* as having "*Corolla* infundibu-
liformis, fauce ampliata . . . *filamenta* basi geniculata, apice
dilatata, *antherae* stigmati incumbentes, apice membrana barbata
terminatae" and *Lochnera* as having "*Corolla* hypocraterimorpha,
fauce contracta . . . *filamenta* brevissima, filiformia, *antherae*
stigmati incumbentes, apice simplices." Thus Endlicher made a
clear distinction between these genera.

Meanwhile George Don, the younger, had also distinguished
these genera in his *General System of Gardening and Botany* (4:95)
retaining the name *Vinca* for the genus containing *V. minor, V.
major* and *V. herbacea* and giving the name *Catharanthus* to the
genus typified by *V. rosea*. Don's protologue is as follows:

XXXVIII. CATHARANTHUS (from καθαροσ, *katharos*, pure,
and ανθοσ, *anthos*, a flower; in reference to the
neatness and beauty of the flowers).-Vinca species,
Lin.
        LIN. SYST. *Pentándria, Monogýnia.* Calyx
5-parted; segments subulate. Corolla salver-
shaped; segments nearly equal sided, obovate,
mucronate; throat bearded; tube long, slender,
clavate at top with 5 tubercles. Stamens in-
closed, conniving over the stigma. Anthers
mucronate, not membranous at top, sessile. Stigma capitate, marginate, bearded at top, and fur-

nished with a cup-shaped membrane below, which
sheaths the upper part of the style. Hypogynous
glands 2, elongated like the ovaria. Follicles
twin, small, terete, glabrous, 2-celled, dehisc-
ing inside; dissepiment double, taking its rise
from the suture, which is plaited inwards. Seeds
16-20 in each follicle, attached longitudinally
to each side of the dissepiment, small, ovate-
acuminated above, grooved and rugged from sharp
tubercles on one side, and smooth on the other
side. Albumen fleshy.-Small shrubs or herbs.
Leaves oppo'site, evergreen, coriaceous. Flowers
elegant, axillary, solitary or twin.
1. C. ROSEUS: downy; branches terete; leaves
elliptic, obtuse, mucronate; petioles bidentate
or bistipulate at the base; flowers axillary,
solitary or twin, sessile. *h.S.* Native every
where within the tropics, but probably originally
from Madagascar. Vinca ròsea, Lin. spec. 305.
Mill. fig. t. 186. Curt. bot. mag. 248. Gaertn.
fruct. 2. p. 172. t. 117. f. 5. Flowers bright
crimson, or peach or rose-coloured, paler on the
under side, with a dark purple eye. Calycine
segments ciliated.
    Var. α, *ròseus;* flowers rose-coloured.
    Var. β, *albus;* flowers white.
    Var. γ, *ocellàtus;* flowers white, with a
purple circle.
    Var. ζ, *villòsus;* leaves villous, rounded
at top, mucronate. Vinca ròsea, Poir. dict. 5.
p. 199.
    Rose-coloured-flowered Catharanthus. Fl.
Feb. Oct. Clt. 1726. Shrub 1 to 2 feet.

This description is on p. 95 of G. Don's 908-page volume 4, and
thus is near its beginning. Publication in parts of volume 4
was announced in Loudon's *Gardener's Magazine* 11:194 (April
1835); p. 23 was cited by his brother David in *Bot. Reg.* 21:
sub t. 1764 (June 1835). Hence the first part issued in 1835
contained at least pp. 1-32 and, assuming that each part contained only that, p. 95 with its description of *Catharanthus*
would have been in the third part, which could well have been
published in 1835. The completed volume was advertised among
"works now first published" in Bent's *Monthly Literary Adver-*

tiser no. 400:37, 48 (10 April 1838), to which I directed attention in *Flora Malesiana* 4:clxxviii (1954), and so must have
been published between 8 March 1838 and 8 April 1838. Thus,
even if *Catharanthus* had been published in the very last part
of G. Don's *General System* vol. 4, it would have priority over
Lochnera Rchb. ex Endlicher (August 1838).

     Lochnera is also invalid as a later homonym. In 1777
Scopoli (*Introductio ad Historiam Naturalem*, p. 271) published
the name *Lochneria* for a genus now united with *Elaeocarpus*.
His protologue is as follows: "1232. LOCHNERIA. Scop. Cal.
4-5 phyllus. Petala quator. Stamina 15-20. Stylus unus.
Bacca unilocularis, monosperma. *Perinkana & Malnaregam* H. MALA-
BAR." Neither he nor Reichenbach indicates the derivation of
their respective names; presumably they both commemorate Michael
Friedrich Lochner (1662-1720) of Nürnberg, who published several
botanical works between 1713 and 1719. Article 75 of the *International Code* states that "when two or more generic names are
so similar that they are likely to be confused . . . they are
to be treated as variants, which are homonyms when they are
based on different types." This article admits some flexibility
of interpretation but the examples given of orthographic variants to be regarded are homonyms, i.e. *Columella* (Vitaceae) and
Columellia (Columelliaceae), both commemorating Columella the
Roman writer on agriculture, and *Eschweilera* (Myrtaceae) and
Eschweileria (Araliaceae), commemorating F. G. Eschweiler (1796-
1831), so exactly parallel *Lochnera* (Apocynaceae) and *Lochneria*
(Tiliaceae) that, even if *Lochnera* Rchb. ex Endlicher (1838)
had had priority over *Catharanthus* (1835-38), it would still
have to be rejected as a later homonym of *Lochneria* Scop.
(1777). The name *Lochneria* had earlier been given by Heister
to a Crucifer grown in the Helmstadt Botanic garden and is mentioned, without being validly published, in Fabricius, *Enumera-
tio methodica Horti Medici Helmstadiensis* 160 (1759), 2nd ed.,
295 (1763).

GENERIC DESCRIPTION

Catharanthus G. Don, Gen. Syst. Gard. Bot. 4:95 (1835).

> *Lochnera* Reichenbach, Consp. Regn. Veg. 1:134
> (1828), nomen nudum; Endlicher, Gen. Pl.
> 583 (1838); non *Lochneria* Scopoli (1777).
> *Vinca* sect. *Lochnera* A. DC. in DC., Prodr.
> 8:380 (1844).
> *Vinca* sect. *Cupa-Veela* A. DC. in DC., Prodr.
> 8:380, 676 (1884).
> *Ammocallis* Small, Fl. S.E. U.S. 935 (1903).

Annual or perennial herbs or small shrubs. Leaves sessile or short-petioled, entire. Flowers terminal or axillary, solitary or in 2-4-flowered cymes, almost sessile or with very short pedicels; bracts absent. Calyx 5-parted, the sepals free almost to the base, narrow, equal, without squamellae on the inside. Corolla small to large, salver-shaped, rose or white; tube cylindric, slender, externally swollen at the insertion of the stamens but contracted at the mouth; lobes 5, broad, spreading, overlapping to the left. Stamens 5, attached to the middle of the corolla tube or just below the mouth, included; filaments very short, not geniculate; anthers free from the stigma, dorsifixed, the connective not prolonged into an apical appendage; pollen ellipsoid or subglobose, smooth, colporate, 25-60 μm., disc represented by two scales much longer than broad, alternating with the carpels. Carpels 2, distinct; ovules numerous (about 10-30), in 2 series in each carpel; style long, slender; clavuncle shortly cylindric, truncate at base. Fruit consisting of 2 long cylindric pointed follicles (mericarps) diverging or parallel. Seeds numerous, small (1.5-3 mm. long), oblong-cylindric, not arillate, with the hilum in a longitudinal depression on one side, blackish, muriculate, the surface minutely reticulate.

Lectotype: *C. roseus* (L.) G. Don (*Vinca rosea* L.)

Distribution: With the exception of *C. pusillus*, an Indian
species, all species of *Catharanthus* are endemic to Madagascar
(Figs. 2,3), but *C. roseus*, introduced from there to Paris and
then from European botanic gardens into the tropics as an orna-

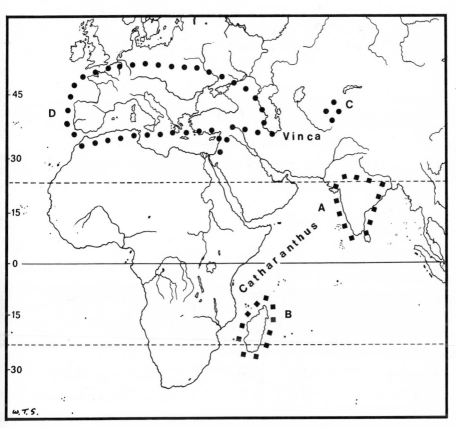

Figure 2.   Distribution of *Catharanthus* and *Vinca*. A,
            Catharanthus pusillus; B, other species of
            Catharanthus before pantropical naturalization
            of *C. roseus*. C, *Vinca erecta*; D, other species
            of *Vinca*.

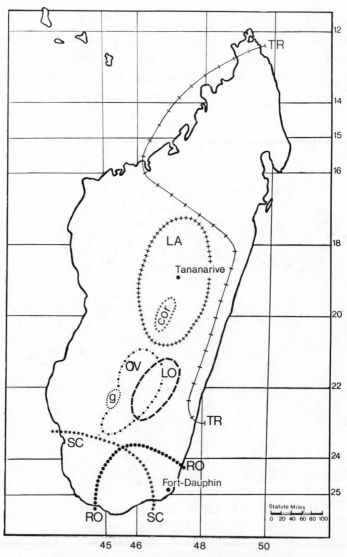

Figure 3. Approximate areas of *Catharanthus* species in
Madagascar: cor, *C. coriaceus*; g, *C. ovalis* ssp.
grandiflorus; LA, *C. lanceus*; LO, *C. longifolius*;
OV, *C. ovalis* ssp. *ovalis*; RO, *C. roseus*; SC,
C. scitulus; TR, *C. trichophyllus*; Fort-Dauphin,
probable type-locality of *C. roseus* (sketch map
based on information from H. Heine and F. Markgraf).

mental plant, has now become pantropical by escaping from cultivation.

Chromosome number: 2n=16 in *C. pusillus, C. longifolius, C.
ovalis, C. roseus* and *C. trichophyllus,* but 2n=32 in the colchicine-induced tetraploids of *C. roseus.*

SECTIONS OF CATHARANTHUS

Stamens inserted about the middle of the corolla tube. Calyx
        not more than 2 mm. long. Follicles not more than
        12 mm. long. Madagascar.

                        Sect. ANDROYELLA (*C. scitulus*)

Stamens inserted just below the mouth of the corolla tube.
        Calyx 3-10 mm. long. Follicles more than 20 mm.
        long:

Corolla tube less than 1 cm. long. Annual. India, Ceylon.

                        Sect. CUPA-VEELA (*C. pusillus*)

Corolla tube more than 1 cm. long. Perennial. Madagascar,
        but one species, (*C. roseus*), now naturalized through-
        out the tropics.   Sect. CATHARANTHUS (6 species)

1.   Sect. CATHARANTHUS
      Vinca sect. *Lochnera* A. DC. in DC., Prodr. 8:380 (1844).
      Catharanthus sect. *Lochnera* (A. DC.) Pichon in Mém. Mus.
          Nat. Hist. Nat. Paris N.S. 27:37 (1949).
      Lochnera sect. *Eulochnera* Pichon, op. cit. 205 (1949) pro
          syn. (cf. op. cit. 237).
      Lectotype: *C. roseus* (L.) G. Don

"Corolla rosea vel alba. Lobi calycis non glandulosi aut sub-
glandulosi. Stamina in superiore parte tubi adfixa, antheris
oblongis, sessilibus. Plantae perennes" (A. DC., loc. cit.).

Leaves often broadest above the middle: apex rounded or obtuse.
        Pantropical.   *C. roseus*

Leaves always broadest at or below the middle; apex acute or
        obtuse. See Fig. 4. Madagascar only:
    Pedicels mostly more than 10 mm. long. Leaves small (not
        more than 30 mm. long, 8 mm. broad).   *C. lanceus*
    Pedicels always short, 10 mm. or less long. Leaves various
        (mostly more than 30 mm. long or 10 mm. broad):
        Corolla tube, follicles pointing downwards, about 15 mm.
            long. Leaf margin revolute.   *C. coriaceus*
        Corolla tube 20-30 mm. long, follicles ascending:
            Leaves not more than 3 times as long as broad,
                mostly ovate or narrowly ovate, sometimes
                lanceolate   *C. ovalis*
            Leaves more than 3 times as long as broad:
                Leaves very narrowly oblong or linear, the
                    base narrowly cuneate   *C. longifolius*
                Leaves mostly lanceolate, rounded or broadly
                    cuneate at base   *C. trichophyllus*

2.   Sect. CUPA-VEELA (A. DC.) Pichon in Mém. Mus. Nat. Hist.
        Nat. Paris N.S. 27:237 (1949).
    Vinca sect. *Cupa-Veela* A. DC. in DC., Prodr. 8:380 (1844).
    Lochnera sect. *Cupa-Veela* (A. DC.) Pichon, op. cit. 205
        (1949) pro syn. (cf. op. cit. 237).
Type (by monotypy): *C. pusillus* (Murray) G. Don.

"Corolla alba, tubo elongato, fauce callosa annulari. Lobi
calycis non glandulosi. Stamina in superiore parte tubi inserta,
filamentis gracilibus, antheris oblongis acutis.-Herba annua"
(A. DC., loc. cit.).

The sectional name is the Malayalam vernacular name for the
only species, *C. pusillus*, i.e. *kapavila* rendered as *cupa-veela*
by Rheede in 1689.

3.   Sect. ANDROYELLA Pichon in Mém. Mus. Nat. Hist. Nat. Paris
        N.S. 27:237 (1949).
    Lochnera sect. *Androyella* Pichon, op. cit. 205 (1949) pro
        syn. (cf. op. cit. 237).
Type (by monotypy): *C. scitulus* (Pichon) Pichon.

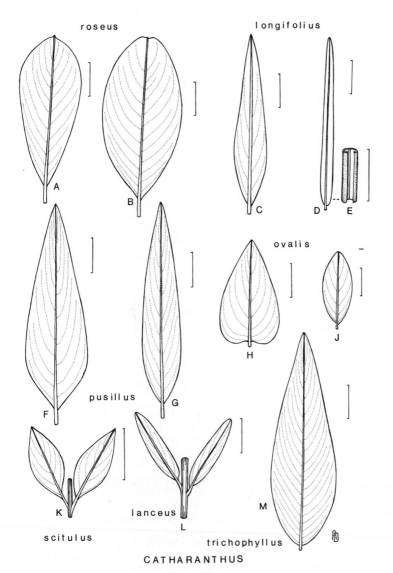

CATHARANTHUS

Figure 4. Leaf shapes in *Catharanthus*.

A–B: *C. roseus* (Curtiss 2270). C–E: *C. longifolius*.
D, upper side; E, lower side (1880, W. Deans Cowan)
F, G; *C. pusillus* (F, Strachey & Winterbottom; G,
cult. Chelsea Physic Garden, 1778). H, J: *C. ovalis*
var. *grandiflorus* (Schlieben 8234). K: *C. scitulus*
(De Cary 3493). L: *C. lanceus* (Hilsenberg & Bojer).
M: *C. trichophyllus* (Schlieben 8035).

The vertical lines represent 1cm. Drawings by
Ann Davis née Farrer (Mrs G. Davis) from specimens
in the British Museum (Natural History), London.

23

"Herbae annuae. Laminae 7-17 (-25) mm. Sepala 1.5-2 mm. Corolla tubus 3.5-5.5 mm., media altitudinae inflatus ac stamini-
fer. Filamenta longiuscula, filiformia. Antherae 0.8-1 mm.
Pollen globosum, 25-30 μm., sulcis inconspicuis. Ovula in
carpello quoque biseriata. Stylus 1.3-2.5 mm. Clavuncula
0.15-0.3 mm., strophio augusto (0.05-0.07 mm.). Mericarpia 6-
12 mm. Semina 1.4-1.8 mm." (Pichon, op. cit. 205).

The sectional name refers to the Androy region of southern Madagascar in which the only species, *C. scitulus*, occurs.

SPECIES OF CATHARANTHUS

Catharanthus coriaceus Markgraf in Adansonia N.S. 10:23 (1970).
    Markgraf's original description of this endemic Madagascar
species is as follows:

              Catharanthus coriaceus Markgraf, *sp. nov.*

        Frutex erectus ad 40 cm altus, modice ramificatus.
    Radix et caudex crassi. Rami striati, inferne folliis
    delapsis cicatricosi. Folia coriacea, glabra, margine
    plerumque revoluta, inferiora ovato-oblonga, obtusa et
    mucronulata, basi rotundata, 20-30 X 8-12 mm, superiora
    linearia, marginibus valde revolutis, 20-30 X 2-3 mm.
    Petiolus 1 mm longus. Flores in axillis foliorum super-
    iorum solitarii. Pedicelli 10 mm longi. Lobi calycis
    lineari-filiformes, acuti, glabri, 8-10 mm longi, vix
    0,5 mm lati. Corollae rubro-violaceae tubus glaber,
    intus ad insertionem staminum pilosus, 15 X 1,5 mm,
    infra faucem constrictum ad 3 mm ampliatus; lobi oblique
    ovati, longiuscule acuminati, glabri, 15 X 5 mm. An-
    therae subsessiles, oblongae, 2 mm longae, apicibus
    acutiusculis faucem attingentes. Ovarium oblongum, bi-
    partitum, 1,5 mm altum. Squamae disci carpellis aequi-
    longae, lineares, acutiusculae. Ovula in utroque car-
    pello biseriata. Caput stigmatis breviter cylindricum,
    basin antherarum attingens. Mericarpia follicularia,

pendentia, cylindrica, acutiuscula, glabra, striata,
30 X 3 mm. Semina fusco-atra, scrobiculata, oblongo-
ellipsoidea, 1,5 X 0,8 mm.

MADAGASCAR, CENTRE: Arivonimamo, à 50 km Ouest
de Tananarive, Mlle Haine 241 (K.). - Betsileo: mon-
tagnes à l'Ouest d'Itremo, bois des pentes occidentales
sur gneiss et quartzites, 1500-1700 m, fl. et fr. 1955,
Humbert 28285 (HOLOTYPE, P); *Humbert 29982.* - Ambato-
finandrahana, pente de l'Ouest, grès, creux de rocher,
fl. 1946, Mlle *Homôlle 1828 et 1884.*
    C. coriaceus is a species of very restricted distri-
bution in central Madagascar and is notable for its coriaceous leaves and its more or less parallel (not diverging) follicles which are directed downwards. It forms a
shrublet to 16in. high with reddish violet flowers.

Catharanthus lanceus (Bojer ex A. DC.) Pichon in Mém. Mus. Nat.

Hist. Nat. Paris N.S. 27:237 (1949).

Vinca lancea Bojer ex A. DC. in DC., Prodr. 8:382 (1844).

Lochnera lancea (Bojer ex A. DC.) K. Schumann in Engler &

Prantl, Natürl. Pflanzenfam. IV. 2:145 (1895).

Tachiadenus parviflorus Baker in Kew Bull. 1897:274 (1897).

Alphonse de Candolle's original description of this endemic

Madagascar species is as follows:

2. V. LANCEA (Boj.! ined.), glabra, caulibus adscen-
dentibus, foliis oblongis subsessilibus basi glan-
dulosis apice mucronulatis obtusis, lobis calycinis
anguste linearibus acuminatis, laciniis corollae
dimidiato-obovatis obtusis mucronulatis tubo duplo
brevioribus. ⚃ in rupestribus provinciae Emirnensis
ins. Madagascar. Caules omnes floriferi, semi-
pedales, simplices vel ramosi, basi sublignosi.
Folia 6-12 lin. longa, 3-4 lin. lata. Pedunculi
pauci, saepius unici ex axilla superiore, foliis
breviores. Lobi calycis 3 lin. longi, basi vix
glandulosi. Corolla rosea? tubo 9-10 lin. longo,
lobis 4-5 lin. longis, patentibus, obtusis, fauce
callosa glanduloso-velutina. Antherae subsessiles,
oblongae, glabrae. Glandulae oblongae. Media inter
V. herbaceam et roseam. (v. s. comm a cl. Boj. et
Bout.)

This is an attractive small-leaved low-growing glabrous
species of *phlox*-like appearance, with many ascending flowering
shoots arising from a perennial woody rootstock. The short
narrowly oblong leaves and comparatively long pedicels distinguish it from other species. It grows among rocks and on dry
slopes of the high plateau in the centre of Madagascar. The
type specimen is at the Conservatoire de Botanique, Geneva;
there are isotypes in the British Museum (Natural History)
and Kew herbaria.

Catharanthus longifolius (Pichon) Pichon in Mém. Mus. Nat.
        Hist. Nat. Paris N.S. 27:237 (1949).

Lochnera longifolia Pichon in Notul. Syst. (Paris) 13:207
        (1948).

Pichon's original description of this endemic Madagascar
species is as follows:

        Lochnera longifolia sp. nov.

        Frutices suffruticesve 40 cm. -1,50 m. alti; foliis
    pubescentibus, petiolo 1,5-3(-7) mm., lamina lineari-
    lanceolata, (25-)40-90 X 3-9 mm., basi longissime
    cuneata; sepalis 3-5 mm. longis; corollae tubo 13-19
    mm. longo, prope apicem staminifero; antheris sessili-
    bus; disci squamis ovario brevioribus; ovulis in pla-
    centa quaque 4-seriatis; pedunculis fructiferis 5-10
    mm., mericarpiis 20-55 mm., seminibus 2-2,5 mm. longis.

        CENTRE-SUD. - Type: *Perrier* 8917.

The label on the type specimen in the Muséum National d'Histoire Naturelle, Paris, reads as follows (fide H. Heine in litt.):
"0.80 à 1 m 50 de haut, rameux. Fl. rose. Vers 950 m. d'al-
titude, cime du Mont Bekinoly près Zazafotsy, bassin du Mank-
goky; gneiss; Septembre 1911."

C. longifolius is a relatively tall-growing species of south-
central Madagascar with stems up to 4½ feet high, leaves to 3½

in. long, the lower ones lanceolate (Fig. 4C), the upper ones linear (Fig. 4D), and rose flowers with a yellow eye.

Catharanthus ovalis Markgraf in Adansonia N.S. 10:23 (1970).

Markgraf's original description of this endemic Madagascar species is as follows:

> *Catharanthus ovalis* Markgraf, *sp. nov.*
>
> Suffrutex ad 40 cm altus. Radix et caudex crassi. Caules herbacei vel sublignosi, erecti vel ascendentes, vix ramificati. Folia sessilia, subcoriacea, ovata, apice obtusa et mucronulata, basi rotundata, 20-25 X 8-12 mm, interdum 30 X 20 mm, glabra vel rarius cum tota planta dense pubescentia. Flores in axillis foliorum superiorum solitarii vel bini. Pedicelli 2-3 mm longi. Lobi calycis lineares, acuti, 4-6 X 0,8 mm. Corollae rubrae vel roseae vel albae tubus 20-30 X 1 mm, infra faucem ad 2,5 mm ampliatus, intus ad insertionem staminum pilosus, extus glaber vel in plantis pilosis velutinus; lobi oblique obovati, breviter acuminati, 10-15 X 6-8 mm aut 20-25 X 12-18 mm. Antherae infra faucem subsessiles, oblongae, obtusae, 2 mm longae, apicibus faucem attingentes. Squamae disci ovario subduplo breviores. Ovarium bipartitum, oblongum, obtusum, glabrum, 2-2,5 mm altum. Ovula in utroque carpello biseriata. Caput stigmatis turbinatum et collo basali cinctum, basin antherarum attingens. Mericarpia follicularia, erecta, cylindrica, acuminata, striata, 30-45 X 3 mm. Semina oblonga, verrucosa, 2,5 X 1 X 1 mm.
>
> Subsp. *ovalis*
>
> Folia saepius 20-25 X 12 mm, glabra vel cum tota planta dense pubescentia. Tubus corollae roseae vel albae 20-30 mm longus, lobi 10-15 X 6-8 mm.
>
> Var. *ovalis*
>
> Tota planta glabra.
>
> MADAGASCAR: répandu aux domaines du Sud-Ouest, Ouest et Centre de 22 à 24° lat. mer. - HOLOTYPE: *Perrier de la Bathie 26538*, P, Sud-Ouest de Betsileo.
>
> Var. *tomentellus* Markgraf, *var. nov.*

Rami, folia, pedicelli, calyces, corolla extus pilis breviusculis vestita.

MADAGASCAR: repandue comme le var. *ovalis*. HOLOTYPE de la variete *tomentellus: Poisson 576,* P, pervenche rose pale, des rochers de l'Isalo, grès, 1500 m, fl. sept. 1922. Cette variété n'a pas d'importance géographique; elle se trouve mêlée à la var. *ovalis*.

Subsp. *grandiflorus* Markgraf, *ssp. nov.*

Folia glabra, saepius late ovata, 30 X 20 mm. Tubus corollae rubrae vel intense roseae 30 mm longus, extus glaber, lobi 20-25 X 12-18 mm.

MADAGASCAR: endémique dans la région de l'Isalo. -HOLOTYPE de la sous-espèce: *Humbert 19585,* P, Ranohira.

The chief distinguishing feature of *C. ovalis* is its usually ovate leaves (Fig. 4H). It forms a shrublet to 16 in. high, with red, rose or white flowers. The specimen illustrated by Farnsworth in *Lloydia* 24: 108, Fig. C3 (1961) as *C. tricho-phyllus* belongs to *C. ovalis*.

Catharanthus pusillus (Murray) G. Don, Gen. Syst. Gard. Bot. 4:
    95 (1835) (See Fig. 5).

Cupa-Veela Rheede, Hort. Ind. Malab. 9:t.33 (1689);
    Cupa-Vela Rheede, op. cit. 61 (1689).

Sinapistrum Indicum diphyllon etc. Plukenet, Phytogr.
    t.119 f.7 (1691).

Papaver corniculatum acre diphyllon Indicum etc., Plukenet,
    Almag. Bot. 280 (1696), fide specim. orig. in Herb.
    Sloane 48, fol. 45.

Vinca pusilla Murray in Nov. Comment. Soc. Reg. Sci. Got-
    ting. 3:66, t.2 f.1 (1773).

Vinca parviflora Retzius, Obs. Bot. 2:14 (1781).

Lochnera pusilla (Murray) K. Schumann in Engler & Prantl,
    Nat. Pflanzenfam. IV. 2:145 (1895).

Figure 5. *Catharanthus pusillus*: engraving of *Vinca pusilla*
reproduced from Nov. Comment. Soc. Reg. Sci.
Gotting. 3: t2 f.1 (1773).

Murray's original description of this widespread Indian species is as follows:

Vinca pusilla MIHI.

VINCA *caule erecto, herbaceo, quadrangulari, floribus geminis solitariisque pedunculatis, foliis lanceolatis.*

Nerium non esse docuit defectus coronae ad faucem. Est igitur haec unica Vincae species herbacea, quum reliquis quatuor, a LINNEO designatis, natura fruticosa competat. Unicum modo exemplum m. Julio procrevit, calore vaporarii adjutum.

　Stirps palmaris.
Radix annua, fibrosa.
Folia seminalia duo, opposita, persistentia, pollicem fere a terra remota, ovato-oblonga, petiolata.
Caulis erectus, strictus, sub foliis seminalibus teres, pubescens, supra quadrangularis, glaber, angulis acutis subulatis, *ramis* in apice paucis, divaricatis, brevibus, pariter quadrangularibus.
Folia lanceolata, integra, acuta, lineata, glabra, patentia, unciam súpraque longa, petiolata, petiolo brevi canaliculato, inferiora opposita, superiora alterna.
Stipulae duae oppositae, patentes, subulatae.
Flores solitarii geminique erecti pedunculati.
Pedunculi teretes, flore triplo breviores, axillares, in superiori caulis et ramorum parte emergentes.
Calyx quinquepartitus, corollae tubo plus quam duplo brevior, persistens, laciniis subulatis, patentibus.
Corolla monopetala, hypocrateriformis, infera, alba, *tubo* infra cylindrico, supra ampliato, *limbo* quinquepartito, plano, laciniis basi angustis, dein ultra medium dilatatis, mucronatis, subaequalibus ore subangulato, tumido, piloso, luteo.
Stamina quinque. *Filamenta* brevissima, erecta, dilatatae tubi parti inserta. *Antherae* oblongae, erectae.
Pistillum unicum, tubo paullo brevius, superum. *Germina* duo oblonga, folliculis duobus membranaceis ovatis obvallata. *Stylus* unus, cylindricus. *Stigma* supra subrotundum, infra vix manifeste in orbiculum dilatatum.
Pericarpium non maturuit.

Murray described this from a plant grown in the Göttingen botanic garden from seed of unknown origin but certainly Indian. Its range extends, mostly in cultivated ground as an unimportant weed, from Garwhal, where it ascends to 5000 ft. (1530 m.), over the Upper Gangetic Plain eastward to Bengal and southward to Madras and northern Ceylon. It is stated to be poisonous to cattle; according to Chopra, Nayar & Chopra, *Supplement to Glossary of Indian Medicinal Plants*, 100 (1969), two alkaloids, pusiline and pusilinine, both heart depressants, and three sterols have been isolated from it.

The following description is based on specimens in the herbaria of the Royal Botanic Gardens, Kew and the British Museum (Natural History):

An annual erect glabrous herb, 10-50 cm. high, much branched from near the base, with four-angled branches. Leaves spreading; blade lanceolate, 1.5-8 cm. long, 0.2-2.5 cm. broad, with the base cuneate, the margin scabrid, the apex acuminate, many-veined, the veins curved and diverging from the midrib at about 25°; petiole 1-10 mm. long. Flowers solitary in leaf axils; pedicels about 1 mm. long. Calyx about 5 mm. long; segments filiform. Corolla white; tube 7-10 mm. long, cylindric, slightly swollen at insertion of stamens shortly below mouth; limb spreading, 6 mm. across, the segments 3 mm. long, 2 mm. broad. Follicles 2-6.5 cm. long; seeds black, longitudinally muriculate, 2.5-3 mm. long, 1 mm. broad.

Catharanthus roseus (L.) G. Don, Gen. Syst. Gard. Bot. 4:95
(1835). See Fig.6.

Vinca foliis oblongo-ovatis integerrimis, tubo floris longissimo, caule ramoso fruticoso Miller, Fig.
Beaut. Pl. 2:124, t.186 (1757).

Vinca rosea L., Syst. Nat. 10th ed., 2:944 (1759).

Lochnera rosea (L.) Spach, Hist. Nat. Vég. Phan. 8:526
(1839).

Figure 6.   *Catharanthus roseus:* engraving of *Vinca foliis*
            oblongo-ovatis from Miller, Fig. Beaut. Pl. 2:
            t 186 (1757), reduced.

Vinca gulielmi waldemarii Klotzsch in Klotzsch & Garcke,
    Bot. Reise Prinzen Ergebn. Waldem. 89, t.70 (1862).

Ammocallis roseus (L.) Small, Fl. S.E. U.S. 935 (1903).

Miller's description of his *Vinca foliis oblongo-ovatis*,
on which Linnaeus based *Vinca rosea*, is as follows:

VINCA *foliis oblongo-ovatis integerrimis, tubo floris
   longissimo, caule ramoso fruticoso.* Periwinkle
   with oblong oval intire Leaves, a very long Tube
   to the Flower, and a branching shrubby Stalk.

The Seeds of this Plant were brought from *Madagascar* to *Paris*, and sown in the King's Garden at *Trianon*,
where they succeeded; and from thence I was furnished
with the Seeds, which succeeded in the *Chelsea* Garden.
It rises with an upright branching Stalk to the Height
of Three or Four Feet; which at first is herbaceous
and succulent, covered with a smooth purplish Skin,
but afterwards it becomes ligneous and tough. This
divides upward into several Branches, which are gar-
nished with oblong oval smooth Leaves, which are
fleshy and intire. At the Base of the Leaves come
out One Flower with a very long Tube; which is divided
at the Top into Five broad obtuse Segments, which
spread open flat, and are of a bright Peach Colour
on their upper Side, as is represented at *a*; but of a
pale Blush Colour on the under, as is shewn at *b*.
These have a very short Empalement, which is cut at
the Brim into Five acute Segments, as is represented
at *c*; and at *d*, the Flower is cut open longitudinally
to shew the Situation of the Five Stamina, which are
closely joined to the Tube of the Petal, so as not
to be separated from it; and the Summits are fastened
to the Mouth of the Tube. *e* represents the short
Empalement which is cut at the Top into Five acute
Segments, in which is situated the Germen supporting
a slender Style; which is the Length of the Tube,
and is crowned with Two Stigmas, shewn at *f*; the
lower is orbicular and compressed, but the upper is
round and concave. These are closely surrounded by
the awl-shaped Summits, One of which is represented
magnified at *g*; and by the Side is another of its
natural Size. *h, i,* and *k,* shew the Empalement
with the Germen, Style, and Stigmas magnified; *l,*
shews the Seed Vessel when ripe cut open longitudinally, and *m,* represents it intire; *n,* shews
the Seeds taken out of the Capsule.

If the Flowers of this Plant are closely smelt
to, they have a faint Odour of the Poppy, but this
is not so strong as to be perceived at a small Dis-
tance. There is a Succession of these Flowers on
the Plants for at least Nine Months in the Year;
for they begin to appear the latter End of *March*
or the Beginning of *April*, and continue to the End
of *December*; so that for Three Months in Winter
they make a fine Appearance in the House.

Miller's account can be supplemented as follows:

An erect short-lived perennial subshrub 30-120 cm. high,
branched from near the base, with four-angled branches, the
stem, leaves, pedicels, calyx, corolla tube and follicles usually puberulent, very rarely glabrous. Leaves spreading; blade
narrowly obovate to narrowly elliptic, 3.5-9 cm. long, 1-3.5
cm. broad, with the base cuneate, the margin ciliate, the apex
rounded or obtuse, sometimes retuse, minutely mucronulate,
many-veined, the veins curved and diverging from the midrib at
about 35° - 45°; petiole 4-10 mm. long. Flowers solitary or
paired in leaf axils or terminal, conspicuous; pedicels about
1 mm. long. Calyx 3-7 mm. long; segments narrowly lanceolate.
Corolla typically with rose limb and greenish tube (f. *roseus*),
sometimes the limb white with pale greenish eye (f. *albus*) or
white with reddish eye (f. *ocellatus*); tube (1.5-) 2.5-3 cm.
long, cylindric, slightly swollen at insertion of stamens shortly below mouth; limb spreading, (1.5-) 2.5-5 cm. across, the
segments broadly obovate, to 2.5 cm. long, 2.2 cm. broad. Follicles 1.5-3.5 cm. long; seeds black, longitudinally muriculate, about 2 mm. long, 1 mm. broad.

Autotetraploids (with 2n = 32) have been raised on several
occasions using colchicine, the first having been made by K.
Furusato at Kyoto in 1938 or 1939 and recorded in *Bot. & Zool.,
Theor. & Applied (Tokyo)* 8:1311 (1940). According to Janaki
Ammal & Bezbaruh in *Proc. Indian Acad. Sci.* B.57:339-342 (1963)
and Dnyansar & Sudhakarar in *Cytologia* 35:237-241 (1970), these
colchicine-induced tetraploids have broader leaves, larger

stomata and larger flowers, than the corresponding diploids
(2n = 16) grown under the same conditions; their pollen sterility, however, is high. Similar tetraploids arising in wild
populations would thus be unlikely to perpetuate themselves
and to survive for long because of their sterility or low fer-
tility.

In the course of its spread and naturalization over the
tropics *Catharanthus roseus* has acquired a diversity of vernacular names: in many English-speaking countries, Madagascar
periwinkle; in the West Indies, old maid, ramgoat rose, Cayenne
jasmine, magdalena, vicaria; in India, Cape periwinkle, church-
yard blossom, deadman's flower, ainskati, billa ganneru, rattan-
jot, sadaphul; in Indonesia, Indische maagdepalm, soldatenbloem,
kembang sari tijna, kembang tembaga; in the Philippines, chi-
chirica; in Japan, nichinchi.

Since the publication of Miller's plate in 1757, this species has been illustrated many times; notably in Curtis's *Bot.
Mag.* 7:t.248 (1793), Iinuma, *Somoku Dzusetsu,* ed. Makino, 2:t.
33 (1907), *Addisonia* 11:t.366 (1926) and Degener, *Fl. Hawaiien-
sis* 2:fam. 305, Catharanthus (1935).

Miller grew his plant in the Chelsea Physic Garden, of
which he was the curator. In accordance with Sir Hans Sloane's
deed of conveyance (cf. *Bot. Soc. Edinburgh Trans.* 41:293-307;
1972) the Garden sent yearly to the Royal Society of London 50
herbarium specimens of plants cultivated the previous year in
the Garden, which are now at the British Museum (Natural History). Among them is a specimen (no. 1849) sent in 1758 under
the name *Vinca foliis oblongo-ovatis integerrimis, tubo floris
longissimo, caule ramoso fruticoso. Miller Icons.,* which evidently came from the same cultivated stock as the plant illustrated by Miller in 1757 and which can accordingly be taken as
the typotype of Linnaeus's *Vinca rosea*. It was probably introduced into cultivation from the French settlement at Fort Dau-

phin in the extreme southeast of Madagascar. In a letter of 18
March 1758 to Linnaeus by David van Royen (1727-1799), director
of the Leiden Botanic Garden, the latter describes a *Nerium e
Madagascar* raised at Leiden from seed given to him by a French
diplomat, which is undoubtedly *C. roseus* and evidently of the
same origin as Miller's *V. rosea*. In a letter of 17 March 1759
D. van Royen states that this *Nerium* is a species of *Vinca*.
Further confirmation of Madagascar as the original home of
C. roseus is provided in a paper by Boiteau, 'Sur la première
mention imprimée et le premier é chantillon de *Catharanthus
roseus* (L.) G. Don', in *Adansonia N.S.* 12: 129-135 (1972).
According to Boiteau, Etienne de Flacourt mentioned it in his
Histoire de la Grande Ile de Madagascar: 130 (1658), under the
Tanosy vernacular name 'Tongue' (now 'tonga'), a plant resembling
Saponaria with a flower like a jasmine either purple or white
and a bitter root; the name 'tonga' is still used for *C. roseus*
by the inhabitants of the region around Fort-Dauphin where
Flacourt resided from 1648-1655 and herbarium species from Mada-
gascar under the name 'Tongue ou Tongha' collected by Flacourt
and illustrated by Boiteau (p.132) confirm this.

Markgraf has described a dwarf variety in *Adansonia* N.S.
12:222 (1972) as follows:

Catharanthus roseus (L.) G. Don var. *nanus* Markgr., var. nov.
Herba vel suffrutex non ulta 15 cm alta, sarmentosa. Folia 1-2
cm. longa, 0,6 - 1cm. lata, elliptica vel etiam suborbicularia,
petiolus 1-2 mm. longus. Sepala brevia, 2,5-3 mm. longa.
Tubus corollae 1.5-2cm. longus, lobi 5x3 mm in flore aperto
non tegentes. Squamae disci ovario longiores.
Type: *Descoings* 1021, Cap Sainte Marie. SUD: *Friedmann* 204,
Tranora-Beloha, *Leandri* 4174.
Catharanthus scitulus (Pichon) Pichon in Mem. Mus. Nat. Hist.
         Nat. Paris N.S. 27:237 (1949).

Lochnera scitula Pichon in Notul. Syst. (Paris) 13:207
(1948).

Pichon's original description of this inconspicuous endemic
Madagascar species is as follows:

Lochnera scitula sp. nov.

Herbae annuae pusillae (3-20 cm.); foliis glabris,
petiolo 0-2(-3,5) mm., lamina oblonga, 7-15(-25) X 2,5-
6,5(-10) mm., basi modice cuneata; sepalis 1,5-2 mm.
longis; corollae tubo 3,5-5,5 mm. longo, medio stamini-
fero; filamentis evolutis, disci squamis ovario brevioribus; ovulis in placenta quaque 2-seriatis; pedunculis
fructiferis 0-2 mm., mericarpiis 6-12 mm., seminibus
1,4-1,8 mm. longis.

SUD-OUEST. Type: *Humbert* 12312.

The label on the type specimen in the Muséum National d'Histoire Naturelle, Paris is as follows (fide H. Heine in litt.):
"De Tsivory à Anadabolava (Mandrare moyen). Etalé sur la terre
nue au gite d'étape d'Anivorano (N.E. de Tsivory). Corolle
blue-violacée. 300-400 m. Décembre 1933."

The low-growing habit and small leaves (Fig. 4K) of this
annual herb native to the dry southern region of Madagascar presumably suggested the epithet *scitulus*, 'neat, trim, elegant',
to Pichon. The flowers are inconspicuous.

Catharanthus trichophyllus (Baker) Pichon in Mém. Mus. Nat.
Hist. Nat. Paris N.S. 27:237 (1949).

Vinca trichophylla Baker in J. Linnean Soc. (London), Bot.
20:204 (1883).

Lochnera trichophylla (Baker) Pichon in Notul. Syst. (Paris)
13:207 (1948).

Baker's original description of this endemic Madagascar
species is as follows:

VINCA (§ LOCHNERA) TRICHOPHYLLA, n. sp.

V. ramulis tetragonis, foliis sessilibus membranaceis oblongis acutis subtiliter pilosis, floribus axil-
laribus solitariis pedicellatis, calycis segmentis
setaceis, corollae tubo cylindrico segmentis oblique
obovatis rubro tinctis, folliculis cylindricis.

A shrub, with slender square glabrous stems.
Stipules cut down to the base into setaceous segments. Leaves in distant decussate pairs, ascending, acute, rounded at the base, 2-3 in. long, finely
hairy on both surfaces, especially beneath. Flowers
solitary in the axils of the leaves, on ascending
pedicels 1/4 to 1/3 in. long. Calyx cut down to the
base into 5 linear-setaceous segments 1/3 in. long.
Corolla with a greenish cylindrical tube an inch
long and 5 oblique obovate segments less than half
as long as the tube. Anthers ovate, 1/12 in. long,
sessile at the glabrous throat of the corolla-tube.
Follicles 2 in. long, slender, cylindrical, arcuate,
marked with close fine vertical ribs.-East coast of
Madagascar, *Baron* 1591 ! Gathered previously by Per-
villé 323 ! 522 ! A near ally of the well-known
V. rosea, L.

Widespread in northern Madagascar and the only species occurring
there, *C. trichophyllus* has also a few isolated more southern
stations on the east coast possibly resulting from accidental
introduction. Its acute leaves (Fig. 4 M) readily distinguish
it from *C. roseus.* It usually has bright red flowers with a
yellow eye and may be up to 1½ feet high. The epithet *trichophyllus* refers to the hairiness of the leaves, not to their
shape, which is usually lanceolate or narrowly ovate. The type
specimen is at the Royal Botanic Gardens, Kew.

CHECKLIST OF CATHARANTHUS AND VINCA

CATHARANTHUS G. Don

Section CATHARANTHUS

1. *C. roseus* (L.) G. Don; Madagascar; naturalized throughout
                    the tropics; 2n=16.

2. *C. ovalis* Markgraf

   subsp. *ovalis*;   Madagascar; 2n=16
   subsp. *grandiflorus* Markgraf; Madagascar;   2n=16

3. *C. trichophyllus* (Baker) Pichon; Madagascar; 2n=16

4. *C. longifolius* (Pichon) Pichon; Madagascar; 2n=16

5. *C. coriaceus* Markgraf; Madagascar.

6. *C. lanceus* (Bojer ex A.D.C.) Pichon; Madagascar.

Section ANDROYELLA

7. *C. scitulus* (Pichon) Pichon; Madagascar.

Section CUPA-VEELA

8. *C. pusillus* (Murray) G.Don; India, Ceylon 2n=16.

VINCA L.

Series MINORES

1. *V. minor* L.; Europe; 2n=46.

Series HERBACEAE

2. *V. herbacea* Waldst. & Kit.; Eastern Europe, Western Asia; 2n=46.

3. *V. erecta* Regel & Schmalh.;   Central Asia.

Series MAJORES

4. *V. difformis* Pourret

   subsp. *difformis*;   Southwest Europe, N.West Africa; 2n=46.

   subsp. *sardoa* Stearn; Sardinia; 2n=46.

5. *V. balcanica* Penzes; Balkan Peninsula.

6. *V. major* L.

   subsp. *major*; southwest Europe; 2n=92.

   subsp. *hirsuta* (Boiss.) Stearn;   Western Asia; 2n=92.

It must be pointed out that the classification of *Vinca* attributed
to me by Aynilian, Farnsworth & Trojanek in Nobel Symposium no 25,
Chemistry in Botanical Classification 189-190 (1974) is of their
devising and not mine; although I am there stated to recognize 16
species belonging to 6 sections; actually, as in the publication of
mine cited, 'Synopsis of the genus *Vinca*' (1973), I have recognized

only 6 species and no sections, the species being too closely
allied and the genus too small to justify sectional division. The
species fall into 3 natural groups as listed above and separated
in the Clavis of my 'Synopsis' (p. 42): series *Minores* (plantae
sempervirentes, calycibus 3–5 mm. longis; typus, *V. minor*),
Herbaceae (plantae herbaceae; typus, *V. herbacea*) and *Majores*
(plantae sempervirentes, calycibus 5–20 mm. longis; typus, *V.
major*). To quote my statement (op. cit. 66), which Aynilian and
others have evidently misunderstood, '*V. herbacea, V. pumila,
V. libanotica, V. sessilifolia, V. bottae, V. mixta and V.
haussknechtii* are here regarded as conspecific. It seems impos-
sible to divide this complex into definable not intergrading taxa.'
This means that, in my opinion, these are *not* to be regarded as
separate species, despite their being listed as such by Aynilian
and others.

MARKGRAF'S KEY TO MADAGASCAR SPECIES

By the courtesy of Prof. Jean F. Leroy and Prof. F. Markgraf
I received late in 1974 a copy of the manuscript on *Catharanthus*
prepared by Markgraf and Pierre Boiteau for the *Flore de
Madagascar*, fam. 169, *Apocynacees,* unfortunately long after the
foregoing synopsis was in the press. This valuable account provides a key, descriptions of the Madagascar species, together with
lists of known localities, by Markgraf and pharmacological,
chemical and cultural observations by Boiteau, who possesses a
unique knowledge of these plants in the wild and in cultivation;
according to Boiteau, *C. longifolius* and *C. roseus* have hybridized
at the Jardin Botanique de Tananarive. Particularly relevant to
pharmacologists is Boiteau's remark that *C. trichophyllus* has been
the subject of much confusion and most of the chemical publications
purporting to this species need to be critically examined with
regard to the identity of the material studied. Thus Gabbai's

1958 work refers not to *C. trichophyllus* but probably to *C. longifolius* and Kim's 1970 work probably to *C. ovalis*.

As an aid to pharmacologists in checking the identity of their material, Markgraf's key (translated by me from the French) is given below by kind permission of Professors Leroy and Markgraf.

1a.   Annual herb with procumbent branches. Leaves
      small, lanceolate. Flowers small, solitary;
      tube of corolla 4 mm. long, equalling the lobes.
      Stamens inserted midway up the tube, with distinct
      filaments. Scales of disc half the length of the
      ovary. Fruits 6–12 mm. long.   1. *C.scitulus*

1b.   Shrublets or perennials. Flowers large. Stamens
      inserted just below the throat, almost sessile.
      Scales of disc exceeding or equalling half the
      length of the ovary. Fruits generally more than
      12 mm. long:
      2a. Stems almost always procumbent–ascending and
           herbaceous, several arising from a big under-
           ground rootstock, rarely erect, up to 40 cm.
           long. Leaves oblong-linear, at the apex
           abruptly rounded, 20–30 mm. long, 4–7 mm. broad.
           Only one flower open at a time on each
           shoot.   2. *C.lanceus*

      2b. Stems erect, ultimately woody at base, often
           with only one stem to a plant. Leaves large-
           er or at least the lower ones broader, 7–20
           mm. broad. Several flowers open at the same
           time on each shoot:

3a. Leaves acute, soft, gradually narrowed
     to the apex. Mouth of the corolla
     yellowish or yellow:

     4a. Plant 30-50 cm. high. Leaves ovate,
     with rounded base, 40-70 mm. long, 15-20
     cm. broad, ciliated on margins and below
     with long scattered hairs. Tube of
     corolla 18-20 mm. long. Scales of disc
     equalling the ovary.   3. *C. trichophyllus*

     4b. Plant 40-150 cm. Leaves lanceolate with
     cuneate base, 40-90 mm. long, 5-9 mm. broad,
     the upper ones linear with strongly revolute
     margins, pubescent. Flowers shorter than the
     upper leaves. Tube of corolla 13-17 mm.
     long. Scales of disc 3/4 the length of
     the ovary.   4. *C. longifolius*

3b. Leaves obtuse, with the apex rounded but mucro-
     nate. Mouth of corolla without yellow blotch:

     5a. Stems with short internodes. Leaves
     coriaceous, the lower ones ovate, the
     upper ones linear. Pedicels 10 mm long.
     Calyx 8-10 mm. long Corolla tube 15 mm
     long. Fruits pendant with mericarps
     not diverging .....   5. *C. coriaceus*

     5b. Stems with longer internodes. Leaves
     all ovate or obovate, little coriaceous.
     Pedicels 2-3 mm. long. Calyx 4-6 mm.
     long. Corolla tube 20-30 mm. long. Fruits

ascending with mericarps recurving and
diverging:

6a. Shrublet about 40 cm high. Leaves
     sessile, firm ovate, 20-30 mm.long.
     Root tape-rooted, stout. Scales of
     disc not equalling half length of
     ovary. Shrublet about 40 cm. high
                              6. *C. ovalis*

     7a. Leaves often broadly ovate
          (length:breadth = 3:2). Corolla
          red or bright rose; tube 30 mm.
          long; lobes 20-25 mm. long
                          subsp. ***grandiflorus***

     7b. Leaves generally narrower (length:
          breadth = 5:2). Corolla rose:
          tube 20-30 mm. long, lobes 10-15
          mm. long   subsp. *ovalis*

6b. Leaves short-stalked, generally
     obovate, rarely elliptic. Root slender.
     Scales of disc more than ½ length of
     ovary......   7. *C. roseus*

     8a. Stems erect, 50-75 cm. high.
          Leaves obovate, 25-65 mm. long.
          Corolla with tube 20-25 mm. long;
          lobes recurving in fully open
          flower. Scales of disc at most
          equalling the ovary ..... var. *roseus*

8b.   Stems ascending at the top, 10-
      15 cm. high. Leaves elliptic,
      sometimes almost orbicular, 10-15
      mm. long. Sepals 2,5-3 mm. long.
      Corolla with tube 15 mm. long;
      lobes not recurving in open flower.
      Scales of disc longer than the
      ovary........   var. *nanus*

Attention should also be called to a paper by Yvonne Veyret,
'Quelques données pour la biosystématique de pervenches malgaches
(genre *Catharanthus* G. Don, Apocynaceae)' in *Candollea* 29: 297-307
(Dec. 1974), which provides another key (p. 301) based on differences of habit and records artificially made hybrids grown at Orsay.

CHAPTER II

THE PHYTOCHEMISTRY AND PHARMACOLOGY OF CATHARANTHUS ROSEUS (L.)

G. DON

Gordon H. Svoboda

Eli Lilly and Company
Indianapolis, Indiana 46206

and

David A. Blake*

Department of Pharmacology and Toxicology
School of Pharmacy, University of Maryland
Baltimore, Maryland 21201

INTRODUCTION

During the planning of a phytochemical screening program
in this laboratory it was recognized that a randomized collection of botanicals from the 250,000 known higher plants would
in all probability be fruitless. Furthermore, the projection
by taxonomists of the existence of some 500,000 species as yet
unidentified made such an approach hopeless. As a result,
attention was focused primarily on the investigation of plants
having reported folkloric usage.

---

*Present address: Departments of Pharmacology
and Gynecology/Obstetrics, School of Medicine, Johns Hopkins
University, Baltimore Maryland.

45

Basing a program solely on randomized folkloric usage can also involve pitfalls. It hardly seems reasonable to assume that the aborigine with his overall primitive culture should posses a sophisticated medical knowledge. Furthermore, many of the plants reportedly used would be classified as panaceas, being useful in countless conditions which could not be duplicated in the animal, a criterion so essential for experimental study. Consequently, a great deal of selectivity had to be exercised, and only those plants would be collected which were useful in the human situation, which conditions could be stimulated in the experimental animal. Furthermore, no preference would be given to the types of compounds contained in the plants, i.e., alkaloids, glycosides, etc. It is the nature of the biological response and pharmacology which determines desirability as a medicinal product - the chemistry is incidental.

Selection of the madagascan periwinkle for inclusion in this phytochemical screening program was made on the basis of its reported folkloric usage as an oral hypoglycemic agent (1,2). ,uent success of two alkaloids isolated therefrom as l oncolytic agents was occasioned by a set of favorable circumstances, along with the introduction of new isolation and purification techniques.

As a result of various extensive phytochemical efforts, some 75 alkaloids have been reported isolated from this plant. While six of these have demonstrable experimental antitumor activity, only two, vincaleukoblastine and leurocristine[1], have found extensive use in the human situation. Three other experimental pharmacological activities, diuretic, hypoglycemic, and antiviral, have been found to be associated with a number of alkaloids obtained from this plant.

---

[1] The United States Adopted Names Committee has approved vinblastine (VLB) and vincristine (VCR) as generic names for these alkaloids.

BOTANICAL CHARACTERISTICS

   Catharanthus roseus (L.) G. Don is an erect, everblooming
pubescent herb or subshrub, one to two feet high. It is pantropical in its occurrence and is cultivated as an ornamental
in gardens throughout the world. Two color varieties, pink and
white, are found in the natural state, while a number of seed
hybrids are commercially available.

   A great deal of confusion regarding the proper nomenclature
of this plant has existed in the past. This periwinkle has
been known variously as *Vinca rosea, Lochnera rosea, Catharanthus roseus* and *Ammocallis rosea*.

   It has been recognized that the genera *Vinca* and *Lochnera*
differ in 34 morphological characteristics and should not be
used as synonyms (3). The name *Lochnera* Rchb. ex Endlicher
(1838) is validly published but must be rejected as a later
homonym of *Lochneria* Scop. (1777) and a synonym of *Catharanthus*
G. Don (1835-1838). Therefore, the correct name for the genus
having *Vinca rosea* as its type is *Catharanthus* G. Don, and the
correct name for the madagascan periwinkle is *Catharanthus roseus* (L.) G. Don (4).

FOLKLORE AND BIOLOGICAL PROPERTIES OF CRUDE EXTRACTS

   While the reported hypoglycemic properties of certain extracts of this plant have never been clinically substantiated,
proprietary preparations have been available in certain parts of
the world (5). (Investigators have examined the properties of this
plant on carbohydrate metabolism with essentially negative results
(6,7). Corkhill and Doutch (8), utilizing an infusion of *C. roseus* leaves in 15 diabetic patients, found no hypoglycemic activity and concluded that the preparation acted as an ideal purgative in those patients suffering from constipation.

   Infusions prepared from the leaves have been reported as
being used in Brazil against hemorrhage and scurvy, as a mouth-

wash for toothache and for the healing and cleansing of chronic
wounds (9).

Chopra and co-workers have reported that the total alkaloids
possess a limited antibiotic activity against *V. cholera* and *M.
pyogenes* var. *aureus*, as well as a significant and sustained hypotensive action (10).

Most recently the *in vitro* activity of selected alkaloids
from *C. roseus* against vaccinia and polio type III viruses has
been reported by Farnsworth and co-workers (11).

(The reported oral hypoglycemic activity of various galeni-
cals prepared from this plant prompted its phytochemical examination in two separate laboratories, independently and unknown
to each other. As with other investigators, neither group substantiated hypoglycemic activity in either normal or experimentally-induced hyperglycemic rabbits. Noble, Beer and Cutts,
however, observed a peripheral granulocytopenia and bone marrow
depression in rats (12,13). Continued investigation led to
their preparation of vincaleukoblastine (VLB) (sulfate), an
alkaloid capable of producing severe leukopenia in rats (14-16).

The observation in our laboratories of experimental oncolytic activity primarily against the P-1534 leukemia, a transplanted acute lymphocytic leukemia, in DBA/2 mice, associated
with certain extracts and fractions thereof (Table I) prompted
an intensive phytochemical research effort for the responsible
principle(s) (17-20). Leurosine, a new dimeric alkaloid closely
chemically related to VLB, was eventually obtained along with
VLB sulfate. The experimental antitumor activity of both of
these alkaloids against the P-1534 leukemia in DBA/2 mice was
first demonstrated in these laboratories.

METHODOLOGY OF EXTRACTION AND PURIFICATION

While the isolation of any alkaloid is an individual problem, several standard techniques exist for the preliminary ex-

TABLE I

Activity of Original Extract of Whole Plant and Crude Fractions Against P-1534 Leukemia

Material	Dosage mg/kg/day	Av. wt. change, g, T/C	Av. survival time, days, T/C	% Increase in survival time [1]	% Indefinite survival
Defatted whole plant extract	120.0	+0.1/+2.7	25.6/14.8	73	0
Fraction A	0.5	-0.6/+1.2	26.5/19.2	38	20
Fraction B	30.0	-1.2/+1.2	24.0/19.0	25	40
Total alkaloids	6.0	+0.8/+2.9	27.7/17.2	61	20
	7.5	+0.2/+0.6	29.8/13.4	122	0
	15.0	-2.3/+0.6	20.3/13.4	51	60
	15.0	+0.3/+0.5	30.0/13.0	130	0
	75 (Oral)	-1.5/+0.6	20.6/13.4	53	0

[1] P-1534 mouse leukemia in DBA/2 mice. Ten mice are intraperitoneally implanted with a suspension of malignant cells and after twenty-four hours five of these receive IP injections of the drug daily for ten days. The maximum tolerated dose is given as predetermined using Swiss white mice. The experiment is designed so the control animals live about 15 days, and a percent prolongation due to drug is calculated. When the mice live three times the life span of the controls (about 45 days) they are considered indefinite survivors, and are rechallenged without treatment to prove their susceptibility to the malignancy.

traction from the crude drug and for the subsequent separation
and purification of the specific alkaloid. Few plants, unfortunately, yield single alkaloidal entities, and the main problem
becomes one involving resolution of complex alkaloidal mixtures
which are overwhelmingly contaminated with non-alkaloidal mate-
rials. *C. roseus* was certainly no exception to this.

Most approaches have failed to take full advantage of the
relative basicities of the alkaloids during extraction, treating
the problem in two steps, isolation and purification. These are,
however, inherently related, and the initiation of purification
procedures as early as possible during extraction can often be
advantageous. Total extraction with ethanol normally unneces-
sarily complicates the subsequent purification which must of
necessity involve the use of methods possessing limited resolving power.

A new technique of selective or differential extraction
was devised for this investigation (Fig. 1) (17, 19). It dif-
fers from the classical approach in that purification is effected
during extraction by virtue of binding the stronger bases in the
crude drug with a solution of a weak organic acid, thereby affording an initial separation of the weak and strong bases.
Final purification of most of the alkaloids was accomplished by
chromatographic separation on Alcoa F-20 alumina, partially
deactivated with 10% acetic acid.

The term "selective" extraction may be a seeming misnomer
because of the great number of alkaloids isolated from *C. roseus*
using this technique. Perhaps the term "differential" would be
more appropriate. However, it must be cited that no one extraction has ever yielded all of these pure compounds. Furthermore, only a relatively few of the alkaloids can be regarded as
major constituents, many being obtained in yields of $1.0 \times 10^{-4}\%$
or less.

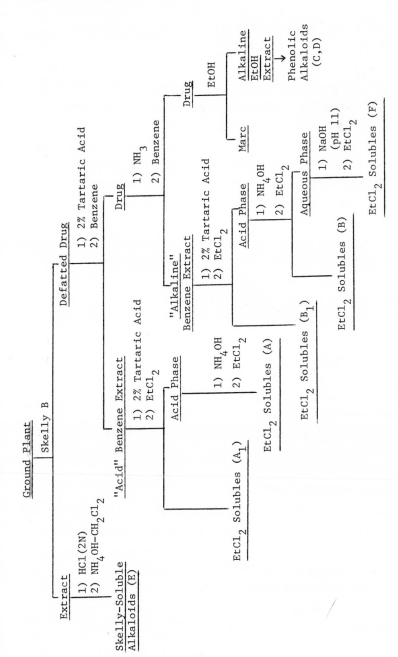

Fig. 1. Extraction scheme.

After the usual defatting process has been completed, the
extraction of the weak bases from the defatted drug is accomplished with benzene, the drug first being intimately mixed
with a 2% aqueous solution of tartaric acid. The tartaric
acid serves to bind the stronger bases in the drug, allowing
the weaker ones to be extracted. In essence, the process allows
for the separation of alkaloids whose tartrates or weak molecular complexes are soluble in benzene from those which are not
under these conditions.

A 2% aqueous solution of tartaric acid is added to the
concentrated benzene extract, and the benzene is removed *in vacuo*.
This process is essentially that of a vacuum steam distillation.
Normally one would resort to using a binary phase extraction,
partitioning the alkaloids between the benzene and an aqueous
acid phase. However, under these conditions the likelihood of
the formation of relatively stable emulsions is real, and obviation of such emulsion formation can be accomplished by this
vacuum steam distillation. The residual acid water usually contains 85% of the alkaloids originally present in the concentrated
benzene extract. The acid-water insoluble material may either
be reworked or discarded as befits the case.

Extraction of the aqueous acid phase at pH 3.0 with ethylene dichloride affords the removal of the very weak bases (A_1
fraction). Extraction of the alkalinized aqueous phase at pH
8.5 with ethylene dichloride yields the A fraction which contains the four major antitumor alkaloids.

After the drug has been alkalinized with ammonia, the
stronger bases can be extracted with benzene. This "alkaline"
benzene extract is worked up in a manner analogous to that for
the "acid" benzene extract, yielding the B_1 and B fractions.
Final extraction of the alkalinized drug with ethanol yields the
strong bases.

Both leurosine and VLB were obtained during a single purification phase (Table II). This separation was indeed fortuitous in that while leurosine was obtained by direct crystallization from methanol, VLB rarely, if ever, crystallized from this solvent. Crystalline VLB sulfate was obtained by direct crystallization from ethanol, whereas leurosine sulfate, formed from the residual base present, did not crystallize from ethanol under the conditions which prevailed.

The early observation that certain fractions produced an unusually high per cent of "indefinite" survivors provided the primary motivation for the comprehensiveness of this entire investigation (Table III). After the isolation of both leurosine and VLB, it was recognized that neither of these alkaloids, nor any therapeutic combination thereof, was responsible for producing the laboratory cures. Subjecting these fractions to any of the known purification techniques proved futile. This challenge eventually led to devising the gradient pH technique (21).

The uniqueness of this technique lies in the fact that alkaloidal separation is achieved under acidic conditions, yielding leurocristine and leurosidine. The gradient pH extraction method utilizes the relative basicities of the alkaloids present in the amorphous starting mixture. The pH of an acid solution of the alkaloids is raised by predefined increments, and the precipitated alkaloids may be isolated by either filtration or extraction with an immiscible solvent. Solvent extraction is to be preferred to filtration as the precipitates which form are usually difficult to filter and may well be prone to oxidation when exposed to warm moist air. The first fractions will contain the less polar alkaloids, the last fractions, the more strongly polar ones.

In actual practice the following method prevails: A 10 g aliquot of the crude fraction is dissolved in 500 ml of benzene.

TABLE II

Chromatography of (A)

Fraction 500 ml ea.	Eluting solvent	Compound	Wt., g	Crystallizing solvent
1	Benzene	Catharanthine	0.250	Methanol
2	Benzene	Vindolinine (as dihydrochloride)	0.210	Methanol–ether
3–19	Benzene	Ajmalicine	0.798	Methanol
20–21	Benzene	Vindoline	0.820	Ether
34–42	Benzene–chloroform (1:1)	Leurosine	0.234	Methanol
43–45	Benzene–chloroform (1:1)	Vincaleukoblastine (as sulfate)	0.126	Ethanol
46	Chloroform	Virosine	0.010	Acetone
47–52	Chloroform–methanol	Amorphous residues	–	–

TABLE III

Anti-P-1534 Activity of Amorphous Fractions Free of VLB and Leurosine

	Dose (mg/kg/day)	Toxic deaths	% prolongation	Indefinite survivors
1.	15.0	5/5	0	0
	6.0	5/5	0	0
	3.0	0/5	?	5/5
2.	3.0	2/5	?	3/5
3.	0.6	2/5	181	0/5
4.	4.5	0/5	151	1/5
5.	9.0	1/5	147	2/5
6.	6.0	0/5	123	4/5
7.	30.0	0/5	238	4/5
8.	6.0	2/5	198	2/5
9.	1.9	0/5	203	4/5
10.	3.8	0/5	85	4/5

Should any benzene-insoluble material be encountered, it is removed by filtration and discarded. The alkaloids are extracted
into 500 ml of 0.1 M citric acid by heating on a steam bath
under reduced pressure ("acid-water run-down" technique). After
removal of all of the benzene, any insoluble material is filtered
off and the filtrate volume is adjusted to 500 ml with H_2O. The
clear acidic solution is then extracted at the existing pH
(usually 2.75-2.85) with 1-500 ml portion of benzene. The pH
of the aqueous phase is subsequently adjusted with NH_4OH solution to 3.40, 3.90, 4.40, 4.90, 5.40, 5.90, 6.40 and 7.50
($\pm$ 0.02), a 500 ml benzene extraction being made at each level.
The benzene solutions, after having been dried over Na_2SO_4, are
concentrated *in vacuo* to dryness. Each amorphous residue is
treated with any of the usual alkaloid crystallizing solvents.
In the case of the *Catharanthus* alkaloids, methanol, ethanol,
acetone and ether appear to be the most appropriate. Occasionally crystalline sulfates can be obtained from the crude mother
liquors and/or from those fractions which did not yield crystalline bases.

Citric acid was chosen to dissolve the alkaloids contained
in the benzene solution as it is a weak, naturally-occurring
acid, and is far less prone to induce changes in labile molecules than any mineral acid. Furthermore, the buffer system
citric acid-ammonium citrate has excellent capacity, thereby
affording a relatively constant pH level during each extraction.

Selective extraction and gradient pH extraction techniques
are by no means universal in scope. They work well with the
indole and dihydroindole monomers and dimers but are apparently
of little value when strong bases are encountered.

The yield of leurocristine is approximately 3×10^{-4}%, this
being the lowest of any medicinally useful alkaloid ever produced
on a commercial basis. Dose-response data for the four major
antitumor alkaloids are presented in Table IV.

The utilization of the new technique of "selective" or
"differential" extraction, coupled with that of the gradient pH
technique, has resulted in the reported isolation of some 64
alkaloids from mature plants (Tables V-IX), 50 of which were first
obtained in these laboratories and three of which were codiscovered in other laboratories. Of the 23 new dimeric alkaloids which
have been isolated, only six, these being indole-indoline in character, have shown oncolytic activity. These are vincaleukoblastine (VLB), leurosine, leurocristine, leurosidine, leurosivine and
rovidine. The activities of the latter two alkaloids are of a
relatively low order of magnitude and do not warrant further discussion at this time. Of these, only VLB and leurocristine are
presently commercially available for the chemotherapeutic management of human neoplasms.[2]

Eleven additional alkaloids, derivatives and glycosides
have recently been discovered as a result of biosynthetic studies
using seedlings or very young *C. roseus* plants. These are tabu-
lated in Table X (51-56).

CHEMISTRY

Preliminary studies of the new active dimeric alkaloids
indicated that they were unrelated to other naturally occurring
antitumor agents such as podophyllin and colchicine, and that
they were indole and dihydroindole in character (57). Correlation of such physical data as electrometric titration values,
ultraviolet, infrared and nuclear magnetic resonance spectra
showed that these alkaloids were closely related chemically and
that they were unsymmetrical dimeric compounds. Microanalyses
of leurosine and VLB and their appropriate salts indicated that
they were $C_{46}H_{56-58}N_4O_9$ compounds (58).

---

[2] VLB is supplied as Velban, Velbe (vinblastine sulfate, Lilly)
and leurocristine is supplied as Oncovin (vincristine sulfate,
Lilly).

TABLE IV

Dose Response Data for Active Alkaloids

Alkaloid (·H₂SO₄)	Dosage mg/kg/day	Toxic deaths	% Increase in survival time[1]	Indefinite survivors
Leurosine	7.5	0	0	0
	10.0	0	32	0
	11.25	0	41	0
	20.0	1	76	1
	150.0 (oral)	0	46	0
Vincaleukoblastine	0.05	0	41	0
	0.10	0	53	0
	0.30	0	70	0
	0.45	0	98	0
	0.45	0	131	0
	0.60	1	150	0
	1.5 (oral)	0	70	0
Leurocristine	0.06	0	24	0
	0.09	0	32	0
	0.12	0	49	0
	0.15	0	55	2
	0.20	1	110	2
	0.25	0	226	3
	0.30	2	?	3
	0.35	2	30	1

Leurosidine			
2.0	0	21	0
3.0	0	30	0
4.0	0	75	0
5.0	0	127	0
7.5	0	?	5
10.0	1	?	4

1   The response of this (Tumour) neoplasm, maintained in the laboratories, to leurosine and VLB has changed during the period of 1958-1962. Leurosine now gives a more consistent activity than formerly (80% prolongation), while VLB currently gives a lower order of activity (50% prolongation) at the maximum tolerated dose. The response to leurocristine and leurosidine during the period of 1960 to the present has remained unchanged.

TABLE V

Alkaloids Previously Reported

Name	Empirical formula	M.P., °C.
Ajmalicine (22)	$C_{21}H_{24}N_2O_3$	253-254
Tetrahydroalstonine (23)	$C_{21}H_{24}N_2O_3$	230-231
Serpentine (22,23)	$C_{21}H_{22}N_2O_3$	156-157
Lochnerine (24)	$C_{20}H_{24}N_2O_2$	202-203
Akuammine[1] (22)	$C_{22}H_{26}N_2O_4$	258-260
Reserpine[1] (25)	$C_{33}H_{40}N_2O_9$	264-265

[1] Not encountered in the studies reported here.

TABLE VI

Monomeric Alkaloids

	Formula	pK'_a	M.P., °C.	Source[2]
Indoles				
1. Alstonine[1] (·HCl) (26)	$C_{21}H_{20}N_2O_3 \cdot HCl$	--	281-282	Rb.
2. Ammorosine (27,28)	----	7.30	221-225	R.
3. Catharanthine (29,28)	$C_{21}H_{24}N_2O_2 \cdot H_2O$	6.8	126-128	L., R.
4. Cathindine (·1/2H₂SO₄) (28)	----	7.25	239-245 (d)	R.
5. Cavincidine (·1/2H₂SO₄) (28)	----	7.85	236-239 (d)	R.
6. Cavincine (·1/2H₂SO₄) (27,28)	$C_{20}H_{24}N_2O_2 \cdot 1/2H_2SO_4 \cdot 1/2H_2O$	6.90	275-277 (d)	L., R.
7. Dihydrositsirikine (30)	$C_{21}H_{28}N_2O_3$	--	215	L., R.
8. Isositsirikine (·1/2H₂SO₄) (31)	$C_{21}H_{26}N_2O_3 \cdot 1/2H_2SO_4$	--	263.5	L., R.
9. Sitsirikine (·1/2H₂SO₄) (32,27)	$C_{21}H_{26}N_2O_3 \cdot 1/2H_2SO_4$	7.6	239-241 (d)	L., R.
10. Vinaspine (33)	----	7.85	235-238	L.
11. Vincarodine (49,78)	$C_{22}H_{26}N_2O_5$	5.9	235-238 (d)	L.
2-Acyl Indoles				
1. Perividine (34)	$C_{20}H_{22}N_2O_4$	neut.	271-279 (d)	L.
2. Perivine (17,19)	$C_{20}H_{24}N_2O_3$	7.5	180-181	L., R.
3. Perosine (·1/2H₂SO₄) (28)	----	7.60	219-225	L., R.
Oxindoles				
1. Mitraphylline (27,35)	$C_{21}H_{26}N_2O_4$	6.20	269-270	L., R.

[1] Not encountered in studies reported here.

[2] Rb, root bark; R, roots; L, leaves.

TABLE VII

Monomeric Alkaloids

	Formula	pK$_a$	M.P., °C.	Source[1]
α- Methylene Indolines				
1. Akuammicine (27,36)	C$_{20}$H$_{22}$N$_2$O$_2$	7.98	181–182	R.
2. Lochnericine (29,37)	C$_{21}$H$_{24}$N$_2$O$_3$	4.2	190–193	L.
3. Lochneridine (32)	C$_{20}$H$_{24}$N$_2$O$_3$	5.5	211–214 (d)	L.
4. Lochnerinine (38)	C$_{22}$H$_{26}$N$_2$O$_4$	––––	168–169	L.
5. Lochnerivine (27,28)	C$_{24}$H$_{28}$N$_2$O$_5$	neutral	278–280	R.
6. Lochrovicine (39)	C$_{20}$H$_{22}$N$_2$O$_3$	4.50	234–238	L.
7. Lochrovidine (39)	C$_{22}$H$_{26}$N$_2$O$_4$	5.60	213–218	L.
8. Lochrovine (39)	C$_{23}$H$_{30}$N$_2$O$_3$	neutral	258–263	L.
Dihydroindoles				
1. Catharosine (40)	C$_{22}$H$_{28}$N$_2$O$_4$	––––	141–143	L.
2. Desacetylvindoline (41,42)	C$_{23}$H$_{30}$N$_2$O$_5$	––––	163–165	L.
3. Maandrosine (·1/2H$_2$SO$_4$) (28)	––––	6.90	160–173	R.
4. Vincolidine (39)	C$_{23}$H$_{26}$N$_2$O$_3$	5.45	165–170	L.
5. Vincoline (39)	C$_{21}$H$_{24}$N$_2$O$_4$	6.1	230–233	L.

6. Vindoline (29,43)	$C_{25}H_{32}N_2O_6$	5.5	154-155	L.
7. Vindolinine (·2HCl) (29,44,79)	$C_{21}H_{24}N_2O_2 \cdot 2HCl$	7.1	210-213 (d)	L.
8. Vindorosine (38,45)	$C_{24}H_{30}N_2O_5$	----	167	L.
Miscellaneous				
1. Ammocalline (27,28)	$C_{19}H_{22}N_2$	7.30	>335 (d)	R.
2. Pericalline (27,28) (Tabernoschizine) (46) (Apparicine) (47) (Gomezine) (48)	$C_{18}H_{20}N_2$	8.05	196-202	R.
3. Perimivine (39)	$C_{21}H_{22}N_2O_4$	indeterminate	292-293 (d)	L.
4. Virosine (17,19)	$C_{22}H_{26}N_2O_4$	5.85	258-261 (d)	R.

1   R, roots; L, leaves

TABLE VIII

Dimeric Indole-Indoline Alkaloids

	Formula	pK'_a	M.P., °C.	Source[1]
1. Carosine (49)	$C_{46}H_{56}N_4O_{10}$	4.4, 5.5	214-218	L.
2. Catharicine (49)	$C_{46}H_{52}N_4O_{10}$	5.3, 6.3	231-234 (d)	L.
3. Catharine (32,75)	$C_{46}H_{54}N_4O_{10}$	5.34	271-275 (d)	L.
4. Desacetyl VLB (·H$_2$SO$_4$) (33)	$C_{44}H_{56}N_4O_8 \cdot H_2SO_4$	5.40, 6.90	>320 (d)	L.
5. Isoleurosine (32)	$C_{46}H_{58}N_4O_8$	4.8, 7.3	202-206 (d)	L.
6. Leurocristine (21,28)	$C_{46}H_{56}N_4O_{10}$	5.0, 7.4	218-220 (d)	L., R.
7. Leurosidine (21,28)	$C_{46}H_{58}N_4O_9$	5.0, 8.8	208-211 (d)	L., R.
8. Leurosine (17,19,28)	$C_{46}H_{56}N_4O_9$	5.5, 7.5	202-205 (d)	L., R.
9. Leurosivine (·H$_2$SO$_4$) (27,28)	$C_{41}H_{54}N_3O_9 \cdot H_2SO_4$	4.80, 5.80	>335 (d)	R.
10. Neoleurocristine (49,75)	$C_{46}H_{54}N_4O_{11}$	4.68	188-196 (d)	L.
11. Neoleurosidine (49,75)	$C_{46}H_{58}N_4O_{10}$	5.1	219-225 (d)	L.
12. Pleurosine (49)	$C_{46}H_{56}N_4O_{10}$	4.4, 5.55	191-194 (d)	L.
13. Rovidine (·H$_2$SO$_4$) (33)	----	4.82, 6.95	>320 (d)	L.
14. Vinaphamine (33)	----	5.15, 7.0	229-235	L.
15. Vincaleukoblastine (14,19,28)	$C_{46}H_{58}N_4O_9 \cdot (C_2H_5)_2O$	5.4, 7.4	201-211	L., R.
16. Vincathicine (33,75)	$C_{46}H_{56}N_4O_9$	5.10, 7.05	>320 (d)	L.

TABLE IX

Miscellaneous Dimeric Alkaloids

	Formula	pK'_a	M.P., °C.	Source[1]
1. Carosidine (49,28)	----	indeterminate	263-278, 283 (d)	L., R.
2. Vincamicine (32)	----	4.80, 5.85	224-228 (d)	L.
3. Vindolicine (32)	$(C_{25}H_{22}N_2O_6)_2$	5.4	248-251 (melts, recryst.) 265-267 (d)	L.
4. Vindolidine (49)	$C_{48}H_{64}N_4O_{10}$	4.7, 5.3	244-250 (d)	L.
5. Vinosidine (27,28)	$C_{44}H_{52}N_4O_{10}$ (Mol.wt. 780)	6.80	253-257 (d)	R.
6. Vinsedicine (50)	(Mol.wt. 780)	4.45, 7.35	206	S.
7. Vinsedine (50)	(Mol.wt. 778)	4.65, 7.0	198-200	S.

[1] L., leaves; R., roots; S., seeds

TABLE X

Alkaloids Isolated from *C. roseus* from Biosynthesis Experiments

Alkaloid	Ref.
N-Acetylvincoside	(51)
Ajmalicine	(52)
Akuammicine	(53)
Catharanthine	(53)
Coronaridine	(53)
Corynantheine	(53)
Corynantheine Aldehyde	(53)
Geissoschizine	(53,55)
Isovincoside	(51)
11-Methoxytabersonine	(53)
Preakuammicine	(53,54)
Stemmadenine	(53)
Tabersonine	(53)
Vincoside	(51,53)
Vindoline	(56)

Comparison of the spectral data for these dimerics with
that of other naturally occurring monomeric alkaloids, particularly catharanthine and vindoline, indicated a close structural
interrelationship between the dimers and these monomers. Furthermore, comparison of the infrared spectrum of an equimolar solution
of catharanthine and vindoline with those of leurosine and VLB
showed them to be virtually imposable from 2-8 μ and extremely
similar up to 16 μ. This observation then allowed the postulation that the dimeric alkaloids leurosine and VLB were composed
of catharanthine- and vindoline-like moieties, with minor mole-

cular modifications, linked together in some unique, heretofore
unknown manner. It then became possible to investigate the
structures of more readily available smaller molecules (I, II)
and to relate this information to the dimeric structures (59,
60).

          I. Catharanthine   II. Vindoline

        Structures for the two clinically active dimeric antitumor
alkaloids, vincaleukoblastine (VLB) (III) and leurocristine (LC)
(IV), have been elucidated by utilizing a combination of chemical
and physical techniques (61), the latter mainly involving the
use of mass spectrometry.

   III.   Vincaleukoblastine, R = CH_3, R' = $COCH_3$
   IV.   Leurocristine, R = CHO, R' = $COCH_3$
   VI.   Desacetyl VLB 4-(N, N-dimethylaminoacetate),
          R = CH_3, R' = $COCH_2N(CH_3)_2$

The complete molecular structure, including the stereochemistry
at C-3,4 and 18 and the absolute configuration of leurocristine
methiodide dihydrate, has been determined by the combination of
two crystallographic methods based on the anomalous scattering
of x-rays (62).

Spectral data (UV, IR and NMR), functional group determination, elemental analysis and mass spectral data for the profoundly active (experimentally) oncolytic alkaloid leurosidine and its
derivatives indicated a formulation of $C_{46}H_{58}N_4O_9$, isomeric with
VLB. The isolation of desacetylvindoline has established the
identity of the dihydroindole portion of the leurosidine moiety,
this being the same as in VLB. The difference between the two
alkaloids therefore resides in the indole portion of the molecules. Subsequent cleavage reactions and studies of the fragmentation patterns have allowed for the postulation of a structure for leurosidine in which the hydroxyl at C-3' is probably
α-oriented (63).

Recent studies involving the use of ^{13}C nmr spectral analysis have positioned the hydroxyl at C-4' in an L-orientation
(75).

Vindoline

V.   Leurosidine, R = $COOCH_3$

The fourth oncolytic alkaloid of interest, leurosine, C_{46}
$H_{56}N_4O_9$, had until recently resisted structural elucidation. A

partial structure has been proposed by Neuss *et al.* (64) but it is at variance with the epoxide structure proposed by Abraham and Farnsworth (65). The work of Wenkert *et al.* utilizing ^{13}C nmr spectral analysis has established the latter proposed structure as being valid (76).

Elucidation of the structures of VLB, leurocristine and leurosidine has allowed for a rational and systematic approach to their synthesis, as well as for a logical pursuit of structure-activity relationships. While these efforts will be detailed in a following chapter, mention of one such derivative can reasonably be made here. In a series of α-aminoacetyl analogs (66), VLB 4-(N,N-dimethylaminoacetate) (VI) displayed excellent experimental oncolytic activity and was selected for clinical testing. While initial results appeared to be favorable (67), certain toxic manifestations involving serious eye problems (corneal and lens changes) were observed in two patients on long-term therapy. Although causal relationship was not definitely documented, clinical trial with this compound has been discontinued (68).

BIOCHEMICAL AND MECHANISM STUDIES

The *in vitro* effect of the four main active oncolytic alkaloids on cells appears to be that of producing metaphase arrest, and this phenomenon has been observed in varying degrees *in vivo* (69). However, differences have been observed with regard to the potency and exact type of action of the various alkaloids. VLB and leurocristine produce a typical C-mitotic effect with micronuclei frequently being observed, while leurosine and leurosidine appear to produce so-called ball-metaphase rather than the classical C-mitotic effect.

The mechanism by which these alkaloids inhibit tumor growth still remains unknown. The above-cited effects can be seen both *in vitro* and *in vivo* in the absence of therapeutic response,

and it is highly unlikely that these cytological effects are
responsible for therapeutic activity. The effects of the various
alkaloids on the incorporation of precursors into DNA, RNA and
protein, as well as the effects on the synthesis of lipids, will
be thoroughly discussed in a subsequent chapter.

BIOLOGICAL TESTING OF CATHARANTHUS ROSEUS ALKALOIDS

The isolation of a profusion of new alkaloids of diverse
chemical structures and the association of oncolytic activity
with a number of these logically dictated an extension into the
investigation of other biological activities.

Table XI summarizes our data on the effects of *C. roseus*
alkaloids in a simple CNS assay to which we refer as the mouse
behavior screen. The methodology employed in this assay has been
previously reported (70). For the most part, the alkaloids act
as CNS depressants, or act on the autonomic nervous system. Pericalline is perhaps the most dramatic in its effects, acting as
a convulsant at doses as low as 2 mg/kg.

A high degree of diuretic activity in saline loaded rats
was established for catharanthine (I) and vindolinine (VII) with

VII. Vindolinine

lesser activity being observed with lochrovicine and vincolidine
(39). A detailed study with other naturally-occurring alkaloids
and derivatives of the active parent compounds established that
the activity was quite specific for the ring systems of the com-

pounds listed in Table XII and to be defined by rigid structural requirements within a given series (71). While structurally unrelated, many of the alkaloids and derivatives possess activities which compare favorably to those of the thiazides. This then represents a new class of compounds which should be investigated either *per se* or via structural modifications to discover new utilizable diuretic agents. A detailed pharmacological study of vindolinine (·2HCl) has been undertaken by Albert and Aurousseau (77).

Our studies have shown that the elusive hypoglycemic activity, which has been disappointingly negative in both experimental and limited clinical investigations, can be attributed to a number of the pure alkaloids (72). Failure in the past to obtain verification of this activity can be attributed to subminimal concentrations of the alkaloids either *per se* or because of the inherent toxicities of the crude extracts or fractions tested.

Catharanthine (·HCl), leurosine (·H_2SO_4), lochnerine, tetrahydroalstonine, vindoline and vindolinine (·2HCl) produce varying degrees of blood-sugar lowering, the onset of which is slow but of a relatively long duration (Tables XIII, XIV). A number of related alkaloids and derivatives have been tested, the corresponding results being given in Table XII.

No physiological correlation or explanation has yet been advanced for the observation that several of the alkaloids and derivatives which exert a hypoglycemic response also produce a significant diuresis.

These alkaloids are all structurally different from the sulfonylureas and therefore represent new leads in this area of research. The use of various galenical preparations of *Catharanthus roseus* as an oral hypoglycemic agent in indigenous medicine cannot be considered to be devoid of merit.

Finally, a large number of *C. roseus* alkaloids were screened for tissue culture anti-viral activity. Of these pericalline was

TABLE XI

Mouse Behavioral Studies with Catharanthus Alkaloids

Alkaloid	Pharmacologic Effect	Dose Range (mg/kg) I.P.	Lethal Dose (mg/kg) I.P.	Comments
Ajmalicine	Muscle relaxation (slight)	100–400	>400	Insignificant activity
Ammorosine	CNS depressant (weak)	200–400	400	Narrow dose range
Carosine	Inactive	400	–	–
Catharanthine (HCl)	CNS depressant, possible autonomic action	25–100	200	Peak – 10 min.; duration – 2¼ hr.
Catharine	CNS depressant (weak)	400	400	Narrow dose range
Desacetyl VLB (sulfate)	CNS depressant	5	5	Toxic
Leurocristine (sulfate)	CNS depressant	1–2.5	2.5–10	Peak – 3 days, slow onset
Leurosidine	CNS depressant & stimulant (mixed)	10–25	50	Peak – 5 min.; duration – 20 min.
Leurosidine (sulfate)	CNS depressant	100	100	Toxic; delayed deaths
Leurosine	Adrenergic blocker	5–25	50–100	–
Lochnerine	CNS depressant & stimulant (mixed)	50–100	200	Adrenergic blocker and convulsant
Lochrovicine	CNS depressant, autonomic act., weak	100–400	>400	Peak – 1 to 2 hr.
Lochrovidine	CNS depressant	100–400	>400	Insignigicaut activity
Lochrovine	Inactive	200	>200	Insufficient quantity for higher doses
Mitraphylline	Inactive	25	–	Insufficient quantity for higher doses

72

TABLE XI (Continued)

Neoleurocristine	CNS depressant	25–100	200	Peak–1 hr.; duration, 4–6 hr.
Pericalline (HCl)	Convulsant	2.5–5	10	Tonic/clonic
Perimivine	Inactive	200	–	Insufficient quantity for higher doses
Perivine (sulfate)	CNS depressant, autonomic activity	10–50	100	Peak–5 min.; duration dose related
Perivine	CNS depressant	50–100	200	–
Sitsirikine (sulfate)	CNS depressant, vasodilator	25–100	200	Peak–15 min.; duration 2–3 hr.
Tetrahydroalstonine	Muscle relaxant (very weak)	100–400	–	Insignificant activity
Vincathicine	CNS depressant	50–200	400	Peak– 9 min.; duration 17 min.
Vincolidine	Inactive	200	–	Insufficient quantity for further testing
Vincoline	CNS depressant	400	400	Insufficient activity
Vindolidine	Inactive	800	–	–
Vindoline	CNS stimulant & depressant	200–1600	400–800	Low safety margin; peak–2 hr.; duration 5 hr.
Vindolinine (2 HCl)	CNS stimulant	50–100	200	Clonic convulsions at 100 mg/kg
Vincaleukoblastine (VLB) (sulfate)				No data available
Vincarodine	Inactive	200	–	Insufficient quantity for further testing

TABLE XII

Diuretic Response to Alkaloids, Derivatives and Thiazides[a]

Name	Dose mg/kg	Urine volume ml	Increase or decrease from control μ-equivalents		
			Na	K	Cl
Ajmalicine	5.0	- 0.85	- 18.5	- 80.5	- 99.0
Ajmaline (·HCl)	50.0	- 0.55	- 220.8*	- 78.8	- 303.5*
	5.0	- 0.60	- 22.2	- 25.2	- 13.2
N-Allylperivinol	50.0	3.40*	429.0*	50.0	445.2*
	5.0	- 2.75	- 116.2	- 158.5	- 70.5
Ammorosine	50.0	- 4.25*	- 177.5*	- 265.0*	- 310.8*
	36.0	- 2.20	- 256.2*	- 114.2	- 311.8*
Carbomethoxy dihydrocleavamine (·HCl)	50.0	1.80	259.5*	47.0*	329.8*
Catharanthine (·HCl)	5.0	0	0	0	-
	50.0	4.20*	326.8*	102.0*	355.5*
Cleavamine (·HCl)	5.0	- 1.45	- 119.0	- 31.8	- 124.2
	50.0	- 2.75*	- 317.8*	- 66.2	- 384.0*
Conopharyngine (·HCl)	5.0	0.73	92.2	103.5	38.5
	25.0	0.30	111.2	54.0	31.2
Coronaridine (·HCl)	5.0	0.53	135.7	17.4	163.2
	50.0	11.71*	1228.4*	302.7*	1627.7*
Descarbomethoxy catharanthine	5.0	- 1.25	- 148.5*	- 52.5*	- 59.2
	50.0	- 2.20*	- 495.8*	- 113.8*	- 575.8*
Dihydrocatharanthine (·HCl)	5.0	1.45	170.0	24.0	99.0
	50.0	All animals died on test			
Dihydrovindolinine (·HCl)	3.8	0.60	192.0	- 12.8	186.2
	37.5	7.85*	1132.8*	111.2	1098.8*

Compound	Dose				
Epi-ibogamine	5.0	− 0.20	31.3	11.8	− 171.5
	50.0	− 1.30	− 133.2	79.0*	− 190.2*
N-Ethyl-0-acetyl-vindolininol	0.9	− 1.35	− 186.5*	90.5*	− 204.0*
	9.0	− 1.60	− 253.2*	0.8	13.2
Ibogamine	5.0	0	49.0	154.2*	134.5
	50.0	9.45*	461.0*	− 254.0	− 222.8
Isoperivine (·HCl)	2.9	− 3.5	− 318.2*	− 160.2	− 50.2
	29.0	− 0.15	62.8	33.0	− 11.8
Isoperivinol (·HCl)	5.0	− 0.10	54.2	95.5	− 289.0*
	50.0	4.35*	− 368.8*	16.8	− 141.8*
Isovindolinine (·2HCl)	5.0	− 2.2*	− 136.2*	221.8	248.2
	50.0	2.2	159.2	2.0	26.0
Isovindolininol	3.75	0.15	− 2.25	91.8	157.2
	37.5	− 1.0	175.0	− 5.6	46.4
Lochnerine	3.6	1.32	− 35.2	− 89.7*	− 268.0*
	36.0	− 2.2 *	− 210.7*	170.4*	− 195.4*
Lochrovicine	3.6	1.77*	146.8*	379.4*	− 708.6*
	36.0	8.32*	634.3*	20.7	− 5.2
Lochrovidine	3.6	− 0.23	65.8	21.7	− 2.7
	36.0	0.82	56.6	− 152.2*	156.8
Mitraphylline	36.0	0.40	195.2	36.8	49.5
N-Methylvindolininol	3.75	0.60	58.0	139.8	186.8
	37.5	2.0	98.0	− 0.8	− 54.0
Perivine sulfate	5.0	− 0.15	27.5	190.5*	− 132.8
	50.0	0	− 168.2*	2.8	83.5
Quebrachidine	5.0	0.80	17.2	− 28.4	− 183.8
	50.0	0.32	− 141.3	− 37.6	− 273.6*
Sitsirikine (·½H₂SO₄)	5.0	1.60*	− 263.1*	− 62.2*	− 344.2*
	50.0	− 2.05*	− 334.5*	− 76.0	116.2
Tabernanthine	5.0	− 2.00	− 124.5	482.8*	− 832.5*
	50.0	13.95*	1076.5*	− 10.0	45.2
Tetrahydroalstonine	5.0	− 0.20	27.8	87.8	60.5
	50.0	− 0.90	44.0		− 273.0*
Vincolidine	36.0	0.85	263.5*	Insignificant from controls	
Vindoline	1.0	− 0.35	99.0	20.2	54.8
	10.0	0.15	56.2	8.8	54.5

TABLE XII (Continued)

Vindolinine	5.0	1.72*	184.0*	53.7	209.3*
Vindolininol	50.0	10.78*	1156.0*	544.8*	1344.2*
Virosine	3.75	−1.93*	−250.3*	22.9	−239.0*
	37.5	−2.25*	−285.5*	49.8	−237.0*
	3.6	−0.97	−44.2	12.1	−103.2
	36.0	−0.97	42.4	13.6	51.7
Voachalotinol	5.0	−0.25	20.8	47.5	−15.0
	50.0	0.70	236.8*	23.8	203.2*
Voacangine (·HCl)	5.0	2.0 *	−40.5	395.8*	47.5
	25.0	11.85*	586.0*	1790.8*	830.0*
Vobasinol (·HCl)	5.0	0.70	111.8	9.2	3.8
	50.0	−0.25	7.2	73.0	−124.2
Chlorothiazide	5.0	1.35*	210.0*	66.1	240.8*
	50.0	3.57*	655.3*	141.2*	666.5*
Dihydrochlorothiazide	0.5	2.90*	390.0*	58.5	444.0*
	5.0	4.55*	861.5*	256.7*	933.8*
Cyclothiazide	0.05	1.65*	337.8*	47.8	345.8*
	0.50	4.96*	799.1*	195.0*	912.8*
	5.0	11.00*	1735.4*	520.8*	1825.7*

a Female rats, weighing 180–250 g and fasted overnight, were orally loaded with normal saline solution, 15 mg/kg. Included in this solution, the test compound was administered as a suspension in 1% acacia. Control animals received only the 1% acacia in saline. The animals were then placed into metabolism cages, and the urine was collected over a 5-hour period. Eight rats, housed two per cage, were used for each dose level. At the end of the test period the urine volume was recorded for each cage. Sodium and potassium were determined on the urines of each cage by flame photometry utilizing an internal lithium standard. Chloride was determined amperometrically by means of a Cotlove chlori-dometer.

* Values significant to the 95% limits of confidence.

Hypoglycemic activity of *C. roseus* alkaloids and standards

Alkaloid	Mean change in blood glucose, mg %					Activity*
	1 hr	2 hrs	3 hrs	5 hrs	7 hrs	
Ajmalicine	+16	+21	+28	+ 4	+15	−
Catharanthine (·HCl)	+20	+ 5	+ 6	−11	−14	+|−
Leurosine (·H_2SO_4)	+ 6	− 7	−22	−28	−31	++
Lochnerine	+ 2	− 3	−10	− 9	−20	+
Perivine	+ 4	+ 4	+14	+11	− 3	−
Sitsirikine (·$\frac{1}{2}H_2SO_4$)	+57	+58	+32	+18	+10	−
Tetrahydroalstonine	+15	+ 3	0	− 8	−17	+|−
Vincathicine (·H_2SO_4)	+ 1	− 3	+12	− 4	+ 2	−
Vindoline	−11	−21	−15	−17	−12	+
Vindolinine (·2HCl)	+11	− 3	−12	−24	−30	++
Acetohexamide[1]	−30	−39	−37	−31	−26 (18 rats)	+++
Tolbutamide[2]	−28	−30	−25	−10	+ 2 (18 rats)	+

*Activity: (−) inactive; (+) questionable; (+) slight; (++) moderate; (+++) strong

[1] N-p-acetylphenylsulfonyl-N′-cyclohexylurea (DYMELOR, Lilly)

[2] N-(Toluenesulfonyl)-N′-n-butylurea (Orinase, Upjohn)

TABLE XIV

Hypoglycemic activity of derivatives and related alkaloids

Alkaloid	Mean change in blood glucose, mg%					Activity*
	1 hr	2 hrs	3 hrs	5 hrs	7 hrs	
Ajmaline (·HCl)	+ 8	+ 4	+ 1	- 2	- 7	-
Allyl perivinol	+15	+11	+ 2	+ 2	-10	-
Cleavamine	+11	+ 6	+11	-10	- 6	-
Coronaridine (·HCl)	+ 9	+11	- 2	-21	-33	+
Desacetylvindoline	+ 6	+ 5	+ 3	-15	-20	$\pm$
Dihydrocatharanthine	+ 8	+ 4	-19	-23	-31	+ (toxic)
Dihydrovindolinine (·2HCl)[1]	- 3	- 2	-10	-21	-20	+
Epi-ibogamine	+27	+23	+15	+27	+11	-
Ibogamine[2]	+ 8	+17	+ 5	- 4	-19	$\pm$[3]
Isovindolinine (·2HCl)[2]	-11	-23	-25	-28	-34	++[3]
Reserpine	+11	+20	+30	+47	+50	-

*Activity: (-) inactive; (±) questionable; (+) slight; (++) moderate

[1] Kindly supplied by Dr. Jean Le Men, l'Ecole Nationale de Medecine et de Pharmacie, Reims (Marne), France.

[2] 50 mg/Kg

[3] Toxic at 100 mg/Kg

the most active, in addition to perivine, leurosivine, leurocristine, vincaleukoblastine, perividine, vindolinine and carosine (73).

CLINICAL MANIFESTATIONS

Both VLB and leurocristine are representatives of a new class of oncolytic agents which are extensively used in the chemo-therapeutic management of a wide variety of human neoplasms. Differences exist in their experimental and clinical tumor spec-tra, and the later will be detailed in a subsequent chapter. The clinical successes obtained with these two alkaloids, leuro-cristine qualifying as a "miracle drug" (74), have given great impetus to the continuing search for tumor inhibitors of plant origin. The clinical uses of these two alkaloids will be dis-cussed in detail in a later chapter.

REFERENCES

1. F. Carcia, "A Botany Symposium on Medicinal Plants", Proceed-ings Eighth Pacific Science Congress of National Research Council of the Philippines, *IVA*, 182 (1954).

2. M. G. Repiton and J. Guillaumin, Bull. Soc. Pharm. Marseilles, 573 (1956).

3. M. Pichon, Mem. Museum Hist. Nat. Paris, XXVII, 153 (1948).

4. W. T. Stearn, Lloydia, 29, 196 (1966).

5. C. T. White, Queensland Agr. J., 23, 143 (1925).

6. L. J. J. Nye and M. E. Fitzgerald, Med. J. Australia 2, 626 (1928).

7. D. H. K. Lee and W. R. M. Drew, *ibid.* 1, 699 (1929).

8. A. B. Corkhill and A. Doutch, *ibid.* 1, 213 (1930).

9. T. Peckholt, Ber. deut. pharm. Ges., 20, 36 (1910).

10. I.C. Chopra, K.S. Jamwal, C.L. Chopra, C.P.N. Nair and P. P. Pillay, Indian J. Med. Research, 47, 39 (1959).

11. N.R. Farnsworth, G.H. Svoboda and R.N. Blomster, J. Pharm. Sci., 57, 2174 (1968).

12. C.T. Beer, British Empire Cancer Campaign, 33rd Annual Report, 487 (1955).

13. J.H. Cutts, C.T. Beer and R.L. Noble, Rev. Canad. Biol., 16, 476 (1957).

14. R.L. Noble, C.T. Beer and J.H. Cutts, Ann. N.Y. Acad. Sci., 76, 882 (1958).

15. R.L. Noble, C.T. Beer and J.H. Cutts, Biochem. Pharmacol., 1, 347 (1958).

16. J.H. Cutts, C.T. Beer and R.L. Noble, Cancer Res., 20, 1023 (1960).

17. G.H. Svoboda, J. Am. Pharm. Assoc., Sci.Ed., 47, 834 (1958).

18. I.S. Johnson, H.F. Wright and G.H. Svoboda, J. Lab. Clin. Med., 54, 830 (1959).

19. G.H. Svoboda, N. Neuss and M. Gorman, J. Am. Pharm. Assoc., Sci. Ed., 48, 659 (1959).

20. I.S. Johnson, H.F. Wright, G.H. Svoboda and J. Vlantis, Cancer Res., 20, 1016 (1960).

21. G.H. Svoboda, Lloydia, 24, 173 (1961).

22. M.M. Janot and J. Le Men, Compt. Rend., 236, 1993 (1953).

23. M. Shimizu and F. Uchimara, J. Pharm. Soc. Japan, 6, 324 (1958).

24. W.B. Mors, P. Zaltzman, J.J. Beerebom, S.C. Pakrashi and C. Djerassi, Chem. Ind. London, 173 (1956).

25. N.K. Basu and B. Sarker, Nature, 181, 552 (1958).

26. P.P. Pillay and T.N. Santha Kumari, J. Sci. Ind. Res., 20B, 458 (1961).

27. G.H. Svoboda, J. Pharm. Sci., 52, 407 (1963).

28. G.H. Svoboda, A.T. Oliver & D.R. Bedwell, Lloydia, 26, 141 (1963).

29. M. Gorman, N. Neuss, G.H. Svoboda, A.J. Barnes, Jr. and N.
     J. Cone, J. Am. Pharm. Assoc., Sci. Ed., 48, 256 (1959).

30. J.P. Kutney and R.T. Brown, Tetrahedron Letters, 1815 (1963).

31. J.P. Kutney and R.T. Brown, Tetrahedron, 22, 321 (1966).

32. G.H. Svoboda, M. Gorman, N. Neuss and A.J. Barnes, Jr., J.
     Pharm. Sci. 50, 409 (1961).

33. G.H. Svoboda, A.J. Barnes, Jr., J. Pharm. Sci., 53, 1227
     (1964).

34. G.H. Svoboda, Lloydia, 26, 243 (1963).

35. L. Michiels, J. Pharm. Belg., 13, 719 (1931).

36. T.A. Henry, J. Chem. Soc., 135, 2759 (1932).

37. C.P.N. Nair and P.P. Pillay, Tetrahedron, 6, 89 (1959).

38. B.K. Moza and J. Trojánek, Chem. Ind. London, 1425 (1962).

39. G.H. Svoboda, M. Gorman and R.H. Tust, Lloydia, 27, 203
     (1964).

40. B.K. Moza and J. Trojánek, Chem. Ind. London, 1260 (1965).

41. D. Gröger and K. Stolle, Naturwissenschaften, 51, 1 (1964).

42. D. Gröger, K. Stolle and C.P. Falshaw, Ibid., 52, 132
     (1965).

43. V.N. Kamat, J. De Sa, A. Vaz, F. Fernandes and S.S. Bhat-
     nagar, Indian J. Med. Research, 46, 588 (1958).

44. M.M. Janot, J. Le Men and C. Fan, Bull. Soc. Chim. France,
     891 (1959).

45. B.K. Moza and J. Trojánek, Collection Czech. Chem. Commun.,
     28, 1419 (1963).

46. U. Renner and P. Kernweisz, Experientia, 19, 244 (1963).

47. B. Gilbert, A.P. Duarte, Y. Nakagawa, J.A. Joule, S.E.
     Flores, J.A. Brissolese, J. Campello, E.P. Carazzoni, R.
     J. Owellen, E.C. Blossey, K.S. Brown, Jr. and C. Djerassi,
     Tetrahedron, 21, 1141 (1965).

48. J.A. Joule, H. Monteiro, L. Durham, B. Gilbert and C. Djerassi, J. Chem. Soc., 4773 (1965).

49. G.H. Svoboda, M. Gorman, A.J. Barnes, Jr. and A.T. Oliver, J. Pharm. Sci., 51, 518 (1962).

50. M. Gorman, G.H. Svoboda and N. Neuss, Proc. Amer. Soc. Pharmacog., Lloydia, 28, 259 (1965).

51. Battersby, A.R., A.R. Burnett, and P.G. Parsons, J. Chem. Soc. C, 1193 (1969).

52. Kutney, J.P., J.F. Beck, V.R. Nelson, K.L. Stuart, and A.K. Bose, J. Am. Chem. Soc., 92, 2174 (1970).

53. Scott, A.I., P.C. Cherry, and A.A. Qureshi, J. Am. Chem. Soc. 91, 4932 (1969).

54. Scott, A.I., and A.A. Qureshi, J. Am. Chem. Soc. 91, 5874 (1969).

55. Battersby, A.R., and E.S. Hall, J. Chem. Soc. D, 793 (1969).

56. Bowman, R.M., and E. Leete, Phytochemistry 8, 1003 (1969).

57. M. Gorman, N. Neuss and G.H. Svoboda, J. Am. Chem. Soc., 81, 4745 (1959).

58. N. Neuss, M. Gorman, G.H. Svoboda, G. Maciak and C.T. Beer, J. Am. Chem. Soc., 81, 4754 (1959).

59. N. Neuss and M. Gorman, Tetrahedron Letters, 206 (1961).

60. M. Gorman, N. Neuss and K. Biemann, J. Am. Chem. Soc., 84, 1058 (1962).

61. N. Neuss, M. Gorman, W. Hargrove, N.J. Cone, K. Biemann, G. Büchi and R.E. Manning, J. Am. Chem. Soc., 86, 1440 (1964).

62. J. W. Moncrief and W.N. Lipscomb, Acta Cryst., 21, 322 (1966).

63. N. Neuss, L.L. Huckstep and N.J. Cone, Tetrahedron Letters, 811 (1967).

64. N. Neuss, M. Gorman, N.J. Cone and L.L. Huckstep, *Ibid.*, 783 (1968).

65. D.J. Abraham and N.R. Farnsworth, J. Pharm. Sci., 58, 694 (1969).

66. W.W. Hargrove, Lloydia, 27, 340 (1964).

67. J.G. Armstrong, R.W. Dyke, P.J. Fouts, J.J. Hawthorne, C. J. Jansen, Jr. and A.M. Peabody, Cancer Res., 27, 221 (1967).

68. C.J. Jansen, Jr., Eli Lilly and Co., personal communication.

69. C.G. Palmer, D. Livengood, A. Warren, P.J. Simpson and I.S. Johnson, Exptl. Cell. Res., 20, 198 (1960).

70. Farnsworth, N.R., L.K. Henry, G.H. Svoboda, R.N. Blomster, M.J. Yates and K.L. Euler, Lloydia, 29, 101 (1966).

71. Gorman, M., R.H. Tust, G.H. Svoboda and J. LeMen, Lloydia, 27, 214 (1964).

72. Svoboda, G.H., M. Gorman and M.A. Root, Lloydia, 27, 361 (1964).

73. Farnsworth, N.R., G.H. Svoboda and R.N. Blomster, J. Pharm. Sci., 2174 (1968).

74. G. Taylor, Cancer Chemotherapy Rept. (Symposium on Vincristine), 52, 453 (1968).

75. Unpublished results.

76. E. Wenkert, D.W. Cochran, E.W. Hagaman, F.M. Schell, N. Neuss, A.S. Katner, P. Potier, C. Kan, M. Plat, M. Koch, H. Mehri, J. Poisson, N. Kunesch and Y. Rolland, J. Am. Chem. Soc., 95, 4990 (1973).

77. O. Albert, Resume of doctoral thesis, "Recherches sur le metabolisme et le mecanisme d'action d'un alcaloide indolique a propriete diuetique: la vindolinine", Universite de Paris (1967).

78. N. Neuss, H.E. Boaz, J.L. Occolowitz, E. Wenkert, F.M. Schell, P. Potier, C. Kan, M.M. Plat and M. Plat, Helv. Chim. Acta, 56, 2660 (1973).

79. A. Ahond, M.-M. Janot, N. Langlois, G. Lukacs, P. Potier, P. Rasoanaivo, M. Sangare, N. Neuss, M. Plat, J. LeMen, E.W. Hagaman and E. Wenkert, J. Am. Chem. Soc., 96, 633 (1974).

CHAPTER III

THE PHYTOCHEMISTRY OF MINOR CATHARANTHUS SPECIES

M. Tin-Wa and N.R. Farnsworth

Department of Pharmacognosy and Pharmacology, College
of Pharmacy, University of Illinois at the Medical
Center, Chicago, Illinois 60612

INTRODUCTION

Interest in certain of the minor *Catharanthus* species in
our laboratories was initiated as a result of reports describing
the isolation of the now clinically useful antitumor alkaloids
vincaleukoblastine (vinblastine, VLB) and leurocristine (vincristine, VCR), in addition to four other active antitumor alkaloids,
namely leurosine (vinleurosine, VLR), leurosidine (vinrosidine,
VRD), leurosivine and rovidine[1], from the pantropical *C. roseus*.
The phytochemistry of *C. roseus* has been covered in Chapter II
of this monograph by Svoboda. Our initial interest in this group
of plants was initiated on chemotaxonomic considerations that
plants of the same genera produce similar or identical chemical
constituents, and hence these minor species would be the logical
plants to explore as new sources of the active antitumor alka-

loids, or structurally similar new active alkaloids. Encourage-
ment for further work came when certain fractions of *C. lanceus,*
C. trichophyllus and *C. pusillus* elicited antitumor and cytotoxic
activities[2-6]. Moreover, the medicinal folkloric reputation of
Catharanthus species posed a challenge to verify the validity of
such uses.

In this chapter, the current status of the phytochemistry
of the minor species of *Catharanthus*, namely *C. lanceus, C. pusillus, C. trichophyllus, C. ovalis* and *C. longifolius* will be
reviewed. The other minor *Catharanthus* species, i.e. *C. coriaceus*
and *C. scitulus* are excluded because no chemical or biological
work on these species has yet been reported. Since the botanical
descriptions, occurrence and pharmacognostic studies on these
species have been described elsewhere in this monograph, we will
confine our remarks to medicinal folklore, biological properties,
non-alkaloids and alkaloid constituents and their distribution.

THE PHYTOCHEMISTRY OF CATHARANTHUS LANCEUS
Folklore and Biological Properties

Catharanthus lanceus leaves have been reported to be used
in South Africa as an astringent[7], bitter[7], and emetic[7], and in
Madagascar the roots have been used as a purgative[8] and vermi-
fuge[9]; the aerial parts as a galactagogue[8] and vomitive[8]; and
the whole plant as a remedy for dysentery. There is no mention
in the literature of *C. lanceus* having been used as an antidia-
betic, for which the related *C. roseus* is reputedly efficaceous[10].

Biological activity studies on extracts and isolates from
C. *lanceus* have been reported by Farnsworth[11], which include hypotensive[2], antitumor[2], hypoglycemic[12] and antiviral effects[13].

Yohimbine, a potent α-adrenergic blocking agent isolated as
a major alkaloid from C. *lanceus* leaves, was shown to significantly reduce blood pressure in normotensive rats and dogs. However, no other hypotensive compounds have been isolated from the
post-yohimbine fraction or other fractions which were also shown
to be markedly hypotensive. Thus, it appears that there is still
at least one highly active hypotensive compound remaining in C.
lanceus[2]. During the same study, pericalline was observed to
elicit a pronounced analeptic effect in test animals[2].

Following submission of a lyophilized aqueous extract of the
roots of C. *lanceus* to the Clayton Biochemical Institute, University of Texas, this extract was found to inhibit the RC mammary carcinoma 70 per cent in mice at doses of 5.0 mg/kg (i.p.).
No apparent evidence of toxicity was noted following autopsy of
the drug-treated animals. Subsequently, larger samples of C.
lanceus leaf and roots were extracted, fractionated and submitted
for testing against the P-1534 leukemia in DBA/2 mice at the Lilly
Research Laboratories and to the National Cancer Institute. All
fractions tested at the former laboratory gave negative results,
but the leaf alkaloid (A) fraction was highly active against
this animal neoplasm at the latter laboratory. This difference
in results has never been explained. Since then we have used

the Eagle's 9KB carcinoma of the nasopharynx in cell culture as
an indicator of cytotoxicity for certain alkaloid fractions.
The criteria for cytotoxicity are such that plant extracts must
exhibit an $ED_{50} \leq$ 15.0 ug/ml, and pure compounds an $ED_{50} \leq$ 1.0
ug/ml in order to be considered as cytotoxic[14]. Cytotoxicity
is not always an effective or reliable means for predicting *in
vivo* antitumor activity, or for determining clinical efficacy.
However, the antitumor alkaloids of *Catharanthus* are all cytotoxic. Thus, the 9KB assay was found to be advantageous in guid-
ing our isolation studies. It has been particularly useful since
it is a rapid assay, and because only small amounts of extract
or pure compound are necessary for testing.

Continued phytochemical examinations yielded the antitumor
and cytotoxic alkaloid leurosine[15], as well as the cytotoxic lochnerinine[16]. Post leurosine and lochnerinine fractions still exhibited antitumor and cytotoxic activities of considerable magnitude suggesting that other phytoconstituents possessing these
activities remain in these fractions and warrant further investigation.

When evaluated for diuretic activity in saline-loaded rats,
and compared with chlorothiazide (+2) and dihydrochlorothiazide
(+3), the following results were obtained with several *C. lanceus*
alkaloids: vindolinine dihydrochloride (+3), catharanthine hydrochloride (+2), vindoline (0), perivine sulfate (0) and tetrahydroalstonine (0). The reduction of the ester moiety of vindoli-

nine caused a reversal of activity and the corresponding alchohol became antidiuretic. Ajmalicine (-1) gave an antidiuretic effect. All alkaloids were administered orally at doses of 5.0 and 50.0 mg/kg except for vindolinine, which was given at 1.0 and 10.0 mg/kg[17,18]. Since the activity of these alkaloids compares favorably with that of the thiazides, which are structurally unrelated, a new class of compounds with potentially utilizable diuretic activity have thus been uncovered.

Several of the *C. lanceus* alkaloids have been examined for hypoglycemic activity in normal rats at a dose level of 100 mg/kg (p.o.). Ajmalicine and perivine were inactive, and catharanthine hydrochloride and tetrahydroalstonine gave questionable results which appeared to be characterized by a delayed onset. Vindoline gave only a slight hypoglycemic effect and leurosine sulfate, as well as vindolinine dihydrochloride were rated as the most active of these alkaloids tested. The hypoglycemic action of the latter two alkaloids was intermediate between that of acetohexamide and tolbutamide[12]. Being structurally different from the sulfonylureas, these alkaloids therefore represent new leads in this area of research.

A routine screening of all available *Catharanthus* alkaloids for tissue culture antiviral activity was quite revealing. Pericalline, perivine, periformyline and vindolinine were found to be inhibitory for the polio type III virus. Perivine, in addition was inhibitory for the vaccinia virus. Alkaloids that were

not inhibitory for either the polio type III or the vaccinia viruses were ajmalicine, leurosine, vinosidine, ammocalline, catharine, lochnerinine, perimivine, tetrahydroalstonine and vincoline[19]. The significance of these findings has not yet been
explored.

The following pharmacological studies were conducted with
3,4-dimethoxyphenylacetamide, isolated from *C. lanceus*: hexo-
barbital sleeping time, anticonvulsant activity, anticholinergic
activity, antidepressant activity, plasma glucose levels, plasma
cholesterol levels and central nervous system (CNS) activity.
Activity could only be detected for the last two parameters. A
significant reduction ($p<0.01$) of plasma cholesterol levels was
observed at the one hour post-injection time in rats receiving
a 100 mg/kg (i.p.) dose. In mice receiving doses of 6.25-200
mg/kg (i.p.), CNS depression was noted, including decreased motor
activity, ataxia and passivity[20].

Non-alkaloid Constituents of Catharanthus lanceus

Choline has been detected in *C. lanceus* by means of paper
chromatography, but it has never been isolated from this plant[21].
Sucrose has been isolated from the roots of *C. lanceus*[22], and
3,4-dimethoxyphenylacetamide was isolated from the leaves[23].
Tannins have also been reported to be present in the plant[7]. A
systematic phytochemical screening of the roots of *C. lanceus* at
the beginning of our own studies revealed the presence of alka-

loids, tannins, saponins and unsaturated sterols, but tests for
flavonoids and cardenolides were negative[24].

Alkaloids from Catharanthus lanceus

Estimates of the quantities of crude alkaloids present in
C. *lanceus* vary from 0.20 to 1.4 per cent in the roots[8,25-27], to
0.10 to 0.15 per cent in the stems[29], and 0.10 to 0.55 per cent
for mixtures of leaves and stems[26-28]. It has been our experience to find 1.36 per cent of crude bases in C. *lanceus* leaves[15]
and 2.10 per cent of crude bases in C. *lanceus* roots[29].

The only phytochemical studies reported prior to 1959 on C.
lanceus were those of Janot and his co-workers, who isolated and
identified ajmalicine (δ-yohimbine)(III)[25,26,30], yohimbine (que-
brachamine)(II)[26,30], tetrahydroalstonine (IV)[30] and lanceine[30]
from the roots (Fig. 1).

In our laboratories we have worked separately on the alkaloids of C. *lanceus* leaves and roots. Major emphasis has been
on the leaf alkaloids since bioassay results have shown antitumor and cytotoxic activity in certain leaf alkaloid fractions.
Since the initiation of investigations on C. *lanceus* in 1959 by
our group, 22 different bases have been isolated, which include
the above four named alkaloids, as well as other alkaloids previously isolated from C. *roseus*, such as leurosine[15,31], lochnerinine[16], perivine[15], vindoline[15], pericalline[29], catharanthine[32], ammocalline[32], vincoline[32], vinosidine[33], catharine[34]
and vindolinine[35].

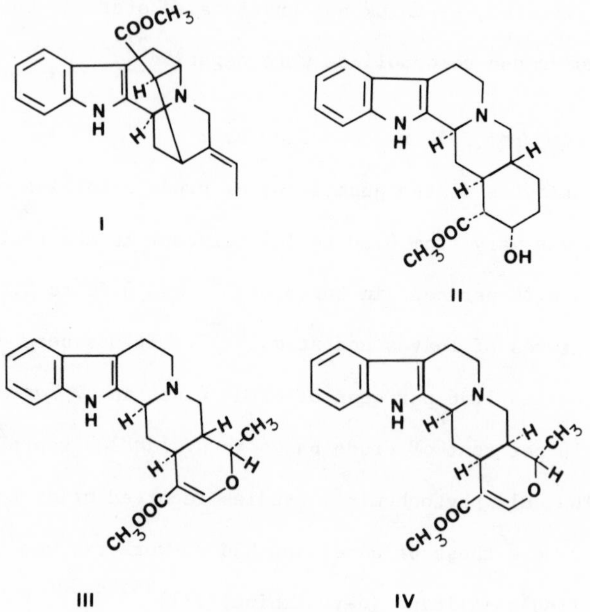

FIG. 1

β-Carboline Alkaloids of *Catharanthus lanceus*.
I, Pericyclivine; II, Yohimbine; III, Ajmalicine,
IV, Tetrahydroalstonine.

Several new bases, such as cathalanceine[29], pericyclivine[36], periformyline[16,37], hörhammericine[38,39], hörhammerinine[38,40] and cathanneine (= cathovaline)[41-43], in addition to a co-discovered alkaloid perimivine[17,29], were discovered by Farnsworth and his co-workers. At about the same time that we reported on the iso-lation, structure elucidation and synthesis of cathanneine, Lang-lois and Potier[44] reported on the isolation and structure eluci-dation of cathovaline from *Catharanthus ovalis*, a newly described *Catharanthus* species. Comparison of the two bases showed them

to be identical and by agreement the name cathovaline should be
retained for this base.

The discovery of these 22 bases was achieved by following
the systematic alkaloid fractionation scheme which was developed
by Svoboda, and which has been a major reason for his success
with C. roseus[45]. We made only minor modifications of Svoboda's
procedures[15]. Each of these fractions from C. lanceus is exceed-
ingly complex. However, most of the mixtures can be separated
well using thin-layer chromatography (TLC) on silica gel G plates
with solvent systems composed of either ethyl acetate-absolute
ethanol (3:1), n-butanol-acetic acid-water (4:1:1) or methanol.

The visualization of Catharanthus alkaloids on TLC plates
has been greatly aided by the use of the ceric ammonium sulfate
(CAS) spray reagent[46], which produces an array of colors with
most indole alkaloids. The color reactions are somewhat associated with the UV chromophores in each alkaloid, but functional
groups on the molecule can cause marked changes in color with
the CAS reagent. However, in most cases, the UV spectrum is
directly related to the colors produced with CAS, e.g. 2-acylin-
doles (Fig. 2), oxindoles and quaternary bases do not react with
the reagent. However, subsequent spraying of TLC plates with
the modified Dragendorff reagent[47] will serve to detect alkaloids
of this class. All β-carbolines (Fig. 1) react to give a yellow
or yellow-green color with CAS, but all α-methyleneindolines
(Fig. 3) give an initial solid blue color with a yellow or orange

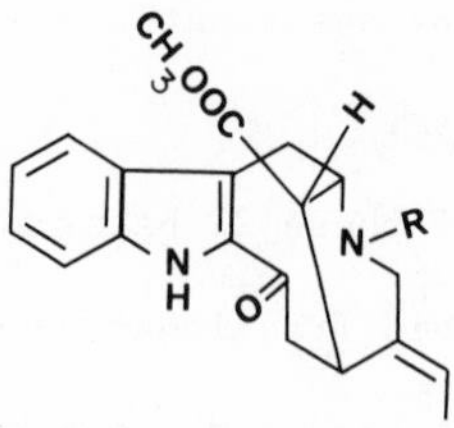

V, VI

FIG. 2

2-Acylindole Alkaloids of *Catharanthus lanceus*.
V, Perivine (R=H); VI, Periformyline (R=CHO).

center quickly forming and expanding in size. All dihydroindole

alkaloids (Fig. 4), or bases having an N-methyl group will give

a red or orange color with the CAS reagent. Alkaloids having

adjacent methoxyl groups on a benzene ring (A ring of indole

alkaloids) will give a light pink color unless they are of a

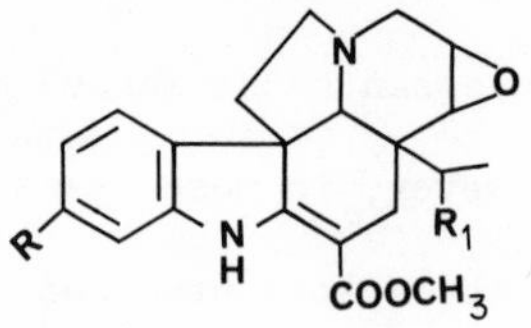

VII, VIII, IX

FIG. 3

α-Methyleneindoline Alkaloids of *Catharanthus
lanceus*. VII, Lochnerinine (R=CH_3O, R_1=H_2);
VIII, Hörhammericine (R=H, R_1=OH); IX, Hörhammerinine (R=OCH_3, R_1=OH).

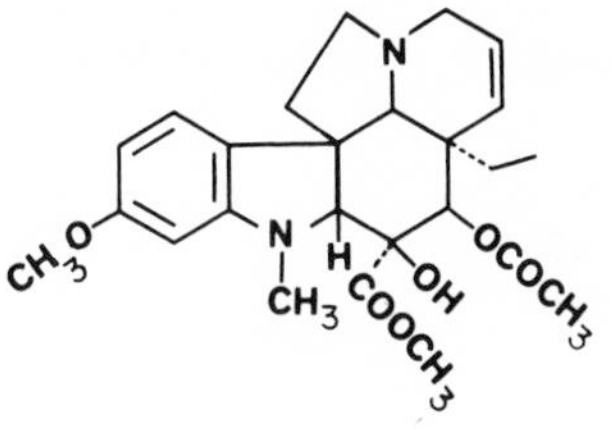

X 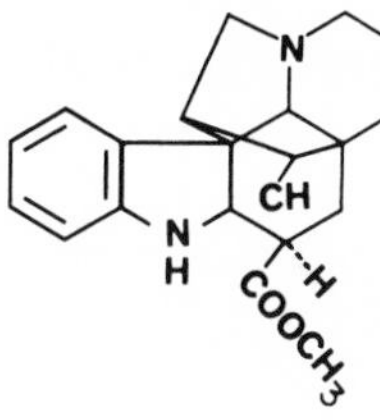   XI

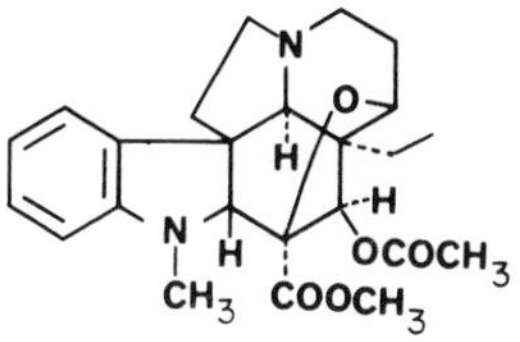

XII

FIG. 4

Dihydroindole Alkaloids of *Catharanthus lanceus*.
X, Vindoline; XI, Vindolinine; XII, Cathovaline
(=Cathanneine).

group which produces a color that covers the pink. We have ob-

served that all α-methyleneindoline bases having an epoxide group

attached to the D ring will give the characteristic blue color

with either a yellow or orange center, which fades after one

hour, but which then gradually develops an emerald-green color.

Lochnerinine (VII), lochnericine, hörhammericine (VIII) and hör-

hammerinine (IX) all contain this epoxy (ether linkage) group

and they give this characteristic reaction. It is interesting

to note that another *Catharanthus* alkaloid having a non-epoxide

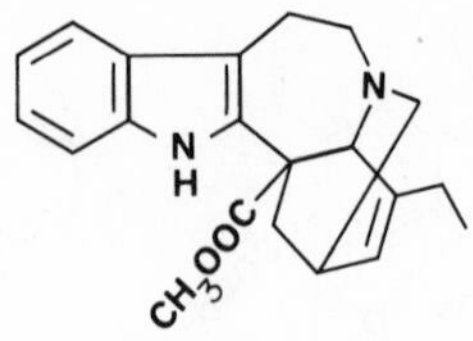

XIII

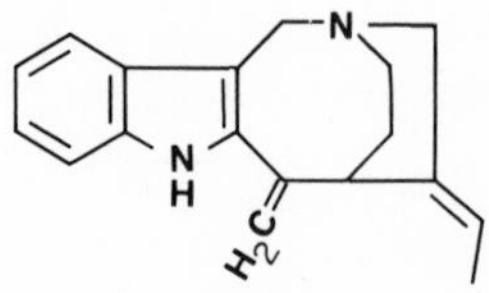

XIV

FIG. 5

Miscellaneous Alkaloids from *Catharanthus lanceus*.
XIII, Catharanthine; XIV, Pericalline.

ether linkage, i.e. cathovaline (XII) gives an initial crimson

color, indicative of the dihydroindole moiety, followed within

one hour by a change to blue with a yellow center, and finally

to a green color in 24 hours. This color change to green appears

to be due to the ether linkage.

By interpretation of chromatograms of the seven crude alka-

loid fractions from *C. lanceus* roots, and the seven fractions

from the leaves, we conservatively estimate that at least 100 dif-

ferent bases are present in this plant. However, we have reason

to believe that many of the minor and trace bases are actually

artifacts.

The dimeric alkaloid leurosine (XV) is extremely labile as

the free base in the dry state or in most organic solvents, and

its decomposition is accelerated by exposure to fluorescent lights
(Fig. 6). We have produced and isolated the major artifact resulting from this decomposition by allowing leurosine to be exposed to fluorescent light in solution, and have also isolated
the same artifact from the leaf alkaloid (A) fraction, which is
rich in leurosine[49].

In our efforts to isolate alkaloids from the 14 leaf and
root crude alkaloid fractions, we initially attempted direct crystallization procedures. Typically, a 5.0 gm sample of the fraction was dissolved in a minimum volume of hot benzene. If insoluble in this solvent, chloroform, ethyl acetate, methanol, ethanol, or acetone were used in that order. The fraction, in solution, was then refrigerated for a minimum period of four days.

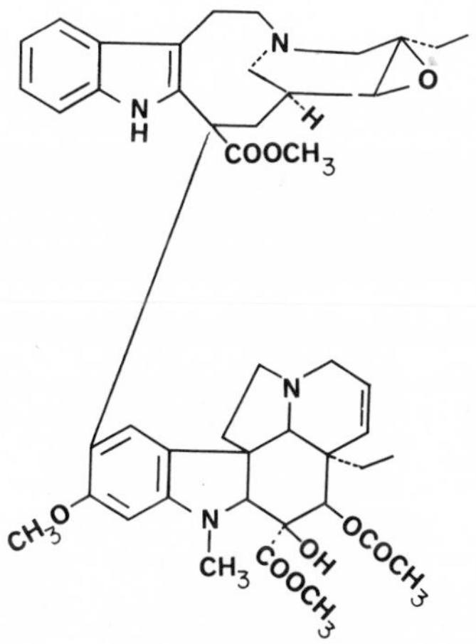

XV

FIG. 6

Structure of Leurosine XV

If no crystallization occurred in that time, the solvent was removed, and the fraction was then dissolved in a minimum volume
of the next solvent in the series, followed by refrigeration etc.
In this manner, we were able to isolate those alkaloids present
in high concentrations in any given fraction, e.g. ajmalicine
(III), yohimbine (II) and perivine (V)[50].

Next, gradient pH separations were performed on the crude
alkaloid fractions in order to obviate tedious column chromatographic technics. Svoboda[51] developed this technique in his work
on *C. roseus*, but he never reported applying it to crude fractions, only to chromatographic cuts from columns. Ajmalicine
was obtained in this manner from the root (A) fraction in a yield
of 15.90 per cent, which was better than the direct crystallization technic (11.85 per cent). Similarly, this same alkaloid
was obtained from the other root fractions in varying amounts
(0.53 to 4.20 per cent). Yohimbine (1.0 per cent) was obtained
from the root (B) fraction although it could not be obtained by
direct crystallization[49].

Gradient pH separation of the leaf (A) fraction yielded 0.20
per cent of ajmalicine, and surprisingly enough, 2.30 per cent
of the active anticancer alkaloid leurosine, which was obtained
previously in our work in a 0.70 per cent yield by means of column chromatography of the same fraction. Comparable yields of
yohimbine (27 per cent) and perivine (20.0 per cent) were obtained from the leaf (B) fractions using gradient pH technics as

with the direct crystallization procedure. These two alkaloids
were also obtained from the leaf (D) fraction where direct crystallization technic produced no results[50].

Thus, by direct crystallization and gradient pH technics on
the 14 crude alkaloid fractions, yohimbine, ajmalicine, perivine
and leurosine could be obtained in yields ranging from 2.30 to
27.0 per cent of the respective crude alkaloid fractions. However, none of the remaining 19 alkaloids that we have isolated
from *C. lanceus* were obtainable by these methods, although interestingly enough, they were obtained in yields of from 0.04 per
cent (vinosidine) to 8.12 per cent (tetrahydroalstonine) (IV) by
column chromatography. Most of these 17 bases were present as
less than 2.0 per cent of their respective alkaloid fractions.

Hence, application of the direct crystallization procedures
or the gradient pH technic appears to be of value in removing
the major alkaloids before resorting to column chromatography to
obtain the minor bases.

As Svoboda found in his work with *C. roseus*, our major success has been realized through the column chromatographic separation of alkaloid mixtures on partially neutralized and partially
deactivated (*ca.* activity grade II or III) alumina. At times,
silicic acid–Celite (5:1) has been used with success, but it does
not have the capacity for resolving complex mixtures as well as
alumina. The eluting solvents usually employed were benzene,
followed by mixtures of benzene and chloroform, chloroform, mix-

tures of chloroform and methanol, and methanol. Thus far we have
been unable to isolate crystalline bases from column fractions
in which methanol was used in admixture with chloroform or by
itself as the eluent. Fractions eluted with methanol usually
contained large amounts of inorganic material, sodium carbonate
and ammonium tartrate being most frequently encountered.

Recently we have obtained success in alkaloid separations
using deactivated silica gel PF_{254}, especially for very purified
fractions[41]. Only one solvent system need be used as the eluent
and the resolution (of alkaloids with Rf's about 0.2 apart and
having Rf values greater than 0.2) found by TLC carries over to
the column chromatographic separation and the development of such
a column with a few grams of purified fractions can be achieved
within a few days. However it was highly unsuccessful with crude
extracts or large fractions and solvent systems of high polarity
such as methanol when the flow eventually ceases.

Following the grouping of chromatographic fractions, direct
crystallizations are made, using the normal organic solvents for
alkaloids. If no crystalline entities are obtained, the fraction
is subjected to the gradient pH technic. Gradient pH fractions
failing to yield pure compounds are converted to sulfate salts
and crystallization attempts are again carried out. It has become quite evident from our work that many of the alkaloids of
C. lanceus exist as oils which are resistant to any crystallization attempts, hence salt formation is necessary to obtain them

in pure form. Vindolinine (XI) is a good example of an alkaloid that cannot be obtained as the free base in crystalline form, but it readily forms a crystalline dihydrochloride.

On the basis of similarities in their basic nucleus, as well as their UV chromophores (Table 1), the 22 *C. lanceus* alkaloids can be classified into six major groups. The structures of 14 of the 22 alkaloids of *C. lanceus* are known and the partial structure for the dimeric catharine has been deduced in our laboratory[34]. However, we do not at present know the structures for ammocalline, lanceine, perimivine, vincoline, vinosidine or cathalanceine. Several of these alkaloids are now being studied in our laboratories.

The physical data for the 22 *C. lanceus* alkaloids are presented in Table 2, and the structures where known are grouped by class in I-XV.

Distribution of Catharanthus lanceus alkaloids

Although many of the *C. lanceus* alkaloids are found in related apocynaceous plants, periformyline (VI), cathalanceine, lanceine, hörhammericine (VIII), hörhammerinine (IX) are restricted to this one species.

Recently, a new indole alkaloid having an ether linkage, i.e. cathanneine was isolated[41]; its structure verified by spectral methods[42] and confirmed by semi-synthesis[43] from *C. lanceus* in our laboratories. Langlois and Potier[44] simultaneously re-

M. Tin-Wa and N. R. Farnsworth

Table 1

Classification and UV Absorption Maxima of Alkaloids
from Minor *Catharanthus* Species

Compound	Species[a]	$\lambda_{max.}$, nm.
I. MONOMERIC ALKALOIDS		
A. <u>β-Carbolines</u>		
Pericyclivine I	L	280, 290
Yohimbine II	L	223, 274, 290(sh)
Ajmalicine III	L,P,T	227, 292
Tetrahydroalstonine IV	L,T	226, 272, 291
B. <u>2-Acylindoles</u>		
Perivine V	L	227, 237(sh), 314
Periformyline VI	L	239, 314
C. <u>α-Methyleneindolines</u>		
Lochnerinine VII	L,P	247, 312, 326
Perimivine	L	232, 302, 340
Hörhammericine VIII	L	299, 327
Hörhammerinine IX	L	245, 325
Cathalanceine	L	226, 296, 323
Lanceine	L	288, 300, 325
D. <u>Dihydroindoles</u>		
Vindoline X	L,P,O	212, 250, 304
Vincoline	L	244, 300
Vindolinine XI[b]	L,T	245, 300
Cathovaline (=Cathanneine) XII	L,O	208, 255, 308
Vindorosine	P,T	250, 302
Cathahelenine	T	207, 246, 297
E. <u>Miscellaneous</u>		
Catharanthine XIII[b,c]	L,O	224, 282, 291
Pericalline XIV	L,T	207, 230(sh), 240(sh), 304
Ammocalline	L	218, 288
Vinosidine	L	226, 254, 259, 300
Coronaridine	O	226, 284, 294

Table 1 (cont.)

Compound	Species[a]	$\lambda_{max.}$, nm.
II. DIMERIC INDOLE ALKALOIDS		
Leurosine XV	L,P	214, 263, 287(sh), 296(sh)
Catharine	L	222, 265, 293, 310(sh)

[a] L, *C. lanceus*; P, *C. pusillus*; T, *C. trichophyllus*; O, *C. ovalis*.

[b] As hydrochloride

[c] Free base has $\lambda_{max.}$ 226, 284, 292 nm.

ported on the isolation of the same base, which they called catho-valine from a newly described *Catharanthus* species, *C. ovalis*.

Vincoline[17], vinosidine[48], ammocalline[52] and catharine[53] also occur in related *C. roseus*. Catharanthine has been obtained from *C. roseus*[54] and from *C. ovalis*. Two *Catharanthus* species, i.e. *C. roseus*[55] and *C. pusillus*[56] have provided the antitumor alkaloid, leurosine. Vindolinine[4,57,58] and perimivine[4,17] have also been found in *C. roseus* and *C. trichophyllus*. Vindoline has also been known to occur only in the other *Catharanthus* species, having been discovered in *C. roseus*[59], *C. pusillus*[5] and *C. ovalis*[44].

Gabunia odoratissima has yielded pericyclivine[60] and peri-vine[60], the latter having also been isolated from *C. roseus*[55].

M. Tin-Wa and N. R. Farnsworth

Table 2

Physical Data of Alkaloids of Minor *Catharanthus* Species[a]

Alkaloid		Formula	m.p. 1°	pK$_a$'	[α]$_D$	Species[k]
Pericalline	XIV	$C_{18}H_{20}N_2$	196-202	8.05[b]		L,T
Ammocalline	-	$C_{19}H_{22}N_2$	>335(d)	7.3[b]		L
Pericyclivine	I	$C_{20}H_{22}N_2O_2$	232-233	6.75[b]	+5.2°[d]	L
Vindolinine	XI	$C_{21}H_{24}N_2O_2$	212-214(d)(HCl) 210-212(d)(HCl)	3.3, 7.1[c]	-8[f]	L,T
Catharanthine	XIII	$C_{21}H_{24}N_2O_2$	179-181(d)(HCl) 126-128(base)	6.8[c]	+65.9°[g] +29.8°[d]	L,O
Coronaridine	-	$C_{21}H_{26}N_2O_2$	225-230	6.1[i]	-10[j]	O
Perivine	V	$C_{20}H_{24}N_2O_3$	181-183	7.6[b]	-121°[d]	L
Lanceine	-	$C_{20}H_{26}N_2O_3$	143-145	-	+64°[e]	L
Ajmalicine	III	$C_{21}H_{24}N_2O_3$	252-254(d)	6.31[c]	-60°[d]	L,P,T
Tetrahydroalstonine	IV	$C_{21}H_{24}N_2O_3$	227-227.5(d)	5.83[c]	-107°[d]	L,T
Cathahelenine	-	$C_{21}H_{26}N_2O_3$	165-165.5(d)	-	-	T
Yohimbine	II	$C_{21}H_{26}N_2O_3$	232-233(d)	7.60[b]	+47°[e]	L
Periformyline	VI	$C_{21}H_{22}N_2O_4$	206-209(d)	>4.0[c]	-	L

Compound	No.	Formula	m.p. (°C)	pK	$[\alpha]$	Species
Perimivine	–	$C_{21}H_{22}N_2O_4$	292-293(d)	insol.	-98.71°[d]	L
Vincoline	–	$C_{21}H_{24}N_2O_4$	222-224(d)	6.1[b]	-37.5°[d]	L
Höhammericine	VIII	$C_{21}H_{26}N_2O_4$	140-144	–	-403°[d]	L
Lochnerinine	VII	$C_{22}H_{26}N_2O_4$	164-166	–	-442°[d]	L,P
Vinosidine	–	$C_{22}H_{26}N_2O_5$	253-257(d) 251-253(d)	6.80[b]	–	L
Höhammerinine	IX	$C_{22}H_{28}N_2O_5$	209.5-211(d)	–	-381°[d]	L
Cathovaline (Cathanneine)	XII	$C_{24}H_{30}N_2O_5$	76-77	–	-73[d]	L,O
Vindorosine	–	$C_{24}H_{30}N_2O_5$	167	4.88[i]	-31°[d]	P,R,T
Vindoline	X	$C_{25}H_{32}N_2O_6$	154-155 153-154	5.5[c] 5.60[b]	+42[d] –	L,P,O
Cathalanceine	–	–	188-190(d)	4.50[b]	–	L
Leurosine	XV	$C_{46}H_{56}N_4O_9$	200-202(d)	4.80,7.10[b]	+59.80[d,h]	L,P
Catharine	–	$C_{46}H_{54}N_4O_{10}$	257-258(d)	–	–	L

[a] Arranged by increasing m.w.; [b] in 33% DMF; [c] in 67% DMF; [d] in chloroform; [e] in ethanol; [f] 2·HCl, in water; [g] in methanol; [h] analysis on the hexahydrate; [i] in methylcellosolve; [j] HCl, in methanol; [k] L, *C. lanceus*, T, *C. trichophyllus*, O, *C. ovalis*, P, *C. pusillus*, R, *C. roseus*.

Lochnerinine is found in *C. roseus*[61], *C. pusillus*[6] and *Vinca herbacea*[62].

Besides *C. roseus*[48,52] and *C. trichophyllus*[4], pericalline (alkaloid E, gomezine, (-)-apparicine, tabernoschizine) has also been found in *Aspidosperma eburneum*[63], *A. gomezianum*[63], *A. multiflorum*[63], *A. pyricollum*[64], *Conopharyngia durissima*[65], *C. holstii*[65], *Schizozygia caffaeoides*[65] and *Vallesia dichotoma*[66].

Tetrahydroalstonine has been reported isolated from *C. roseus*[67] and *C. trichophyllus*[3,4] as well as from *Alstonia constricta*[68], *Rauvolfia ligustrina*[69], *R. sellowii*[70,71], *R. vomitoria*[72] and *Uncaria gambir*[73].

In addition to being found in *C. roseus*[74], *C. pusillus*[5,6] and *C. trichophyllus*[4], ajmalicine has been found in such other apocynaceous species as *Rauvolfia serpentina*[75], *R. verticillata*[76], *R. vomitoria*[77], *R. yunnanensis*[78], *R. caffra*[79], *R. canescens*[80-82], *R. sellowii*[71,83], *R. chinensis*[84], *R. heterophylla*[85], *R. tetraphylla*[86], *R. micrantha*[87], *Stemmadenia obovata*[88], as well as the rubiaceous *Mitragyna javanica*[89], *M. speciosa*[90] and *M. javanica* var. *microphylla*[91] and *Pausinystalia yohimbe*[92].

Although yohimbine has been isolated from only one *Catharanthus* species, i.e. *C. lanceus*, it is known to be present in other apocynaceous plants, i.e. *Aspidosperma discolor*[93], *A. excelsum*[94], *A. oblongum*[95], *A. quebracho-blanco*[96], *A. pyricollum*[64], *Corynanthe paniculata*[92], *C. johimbe*[97], *C. macroceras*[98], *C. yohim-*

bi[99], *Pausinystalia yohimbe*[92], *P. angolensis*[92], *Rauvolfia verticillata*[76], *R. serpentina*[100,101], *R. amsoniaefolia*[102], *R. suma-
trana*[103], *R. fruticosa*[103], *R. vomitoria*[104], *R. canescens*[105],
R. heterophylla[85,106], *Ladenbergia hexandra*[107], *Pouteria* sp.[108,109]
and *Diplorrhynchus condylocarpon* ssp. *mossambicensis*[110].

THE PHYTOCHEMISTRY OF CATHARANTHUS PUSILLUS

Folklore and Biological Properties

In India, *C. pusillus* is boiled in oil and rubbed on the
loins for lumbago[111]. It has also been cited as toxic for cattle, and as a cardiac poison[112].

Two of the seven alkaloid fractions prepared from *C. pusillus* whole plant in our laboratories[15] have been shown to elicit
moderate to marked hypotensive activity in rats[5]. Of six *C. pusillus* leaf alkaloid fractions tested for cytotoxic activity,
three were found to be active. The cytotoxic alkaloid lochnerinine was isolated from one of these fractions and the cytotoxic,
antitumor alkaloid leurosine was isolated from another fraction.

Non-Alkaloid Constituents of Catharanthus pusillus

A neutral substance, identified by Battersby and Kapil[113]
as N-benzoyl-L-phenylalaninol, was isolated from a weakly basic
alkaloid fraction of *C. pusillus* in a yield of 1 X 10^{-4} per cent.
In addition, three unidentified sterols[114] and ursolic acid have
been isolated from this plant[56].

Alkaloids from Catharanthus pusillus

Prior to our investigations of this plant, two poorly characterized amorphous alkaloids, pusiline and pusilinine, each
possessing cardiotoxic properties, were reported as being isolated by Majumdar and Paul[114]. We have found that there are 0.6
per cent of crude alkaloids in the leaves of *C. pusillus*[6] and
0.49 per cent in the whole plant[5].

Preliminary examination of this species in our laboratories[5]
yielded the alkaloids ajmalicine and vindorosine. Vindoline was
presumed present by TLC examination. The isolation method employed was essentially the same as that employed for *C. lanceus*,
and the two abovenamed alkaloids were isolated from the gradient
pH separation of fractions A and A_1 respectively. Vindoline was
present in the latter fraction.

In a different examination of *C. pusillus* leaves, Tin-Wa
et al.[6,56] utilized a different type of fractionation procedure
(Scheme I) than had been used previously in our laboratories for
Catharanthus species. Column chromatography of fraction I, using
deactivated Alcoa F-20 alumina, resulted in the isolation of the
cytotoxic alkaloid lochnerinine as the hydriodide salt. Fraction
III, representing the non-phenolic tertiary bases, was subjected
to the gradient pH separation into five groups as follows:

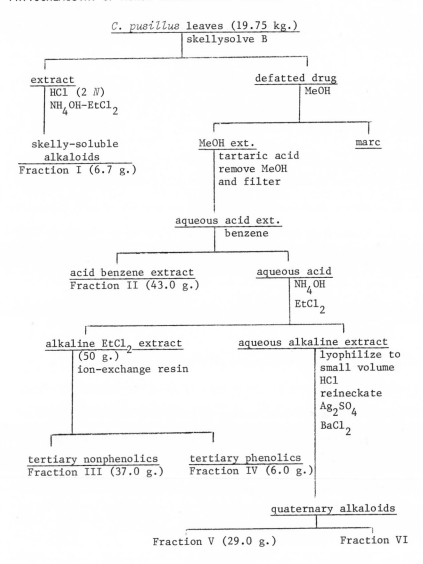

SCHEME I

Fractionation procedure for *C. pusillus*.

M. Tin-Wa and N. R. Farnsworth

Group	pH	Fraction wt.(g)
III-A	2.7 - 3.2	3.42
III-B	3.7	1.84
III-C	4.2 - 5.2	3.55
III-D	5.7 - 7.2	4.73
III-E	7.9 - 9.2	1.25

Direct crystallization of group III-A and III-B provided crystalline vindorosine (0.2%) and ajmalicine (0.04%) respectively. Neither of these alkaloids was found to be cytotoxic and the active entity in fraction III remains unknown. Hence, column chromatographic separation of fractions III-B and III'-B was undertaken[56], using deactivated Alcoa F-20 alumina, which resulted in the isolation of leurosine, the antitumor and cytotoxic alkaloid. Part of the hypotensive activity noted earlier[5] for this plant may now be attributed to this base, which has been shown to elicit transient hypotensive activity[1]. The remaining sub-fractions of fraction III were complex and have not yielded alkaloids by conventional crystallization technics. Further purification work is in progress.

Distribution of Catharanthus pusillus Alkaloids

Of the five well-defined alkaloids isolated from *C. pusillus*, i.e. lochnerinine[6], vindorosine[5,6], ajmalicine[5,6], vindoline[5] and leurosine[56], all are also found in *C. lanceus* and *C. roseus*, as well as other species of plants, which have been previously discussed.

THE PHYTOCHEMISTRY OF CATHARANTHUS TRICHOPHYLLUS

Folklore and Biological Properties

There is no reported medicinal folklore associated with this species.

Seven alkaloid fractions of *C. trichophyllus* aerial parts, prepared as previously described for *C. lanceus*[15], were found to have antitumor activity associated with the A and B_1 fractions[3], while the A and A_1 fractions were cytotoxic. In a subsequent study in our laboratory a different alkaloid extraction scheme was utilized for this plant[5,15]. The total crude alkaloid fraction was found to be active against the Eagle's 9KB carcinoma (cytotoxic), whereas another fraction theoretically devoid of alkaloids was shown to be active against the P-388 leukemia.

Non-Alkaloid Constituents of Catharanthus trichophyllus

Ursolic acid is the only non-alkaloid constituent to be reported isolated from this species[3,4]. During a routine phytochemical screening[4], alkaloids, unsaturated sterols and triterpenes were detected in this plant, and anthraquinones, cardenolides, coumarins, cyanogenic glycosides, flavonoids, leucoanthocyanins and saponins were not detected.

Alkaloids from Catharanthus trichophyllus

Prior to our investigations on this plant, the only phytochemical report in the literature for *C. trichophyllus* was the isolation of ajmalicine from the aerial parts by Gabbai[115]. Ex-

amination of the aerial parts of this species in a manner anala-
gous to our earlier work on *C. lanceus*, resulted in the isolation
of tetrahydroalstonine and vindorosine[3]. Column chromatography
of fraction A on deactivated Alcoa F-20 alumina yielded the alkaloid vindorosine and ursolic acid. In addition to tetrahydroalstonine, ursolic acid was again isolated from fraction B_1 after
passing this fraction over a column of silicic acid-Celite (5:1).

In a later study, a different fractionation scheme was employed[4], in which the plant was initially defatted with Skelly-
solve B, which yielded ursolic acid. The defatted marc was then
exhaustively extracted with methanol. The total crude alkaloid
fraction, which was shown to have cytotoxic activity of a low
order, was processed utilizing the usual acid-base shakeout treatment of the methanol extract. After converting the total crude
alkaloid fraction into citrate salts, the fraction was further
purified by dialysis through a cellophane membrane according to
the method of Klasek[116]. Recovery of the alkaloids was effected
by basification of the dialysate, followed by extraction with
benzene. The partially purified fraction thus obtained was chromatographed on Sephadex LH-20, using methanol-ethylene dichloride
(7:3) in order to separate the high molecular weight alkaloids
(dimers) from the low molecular weight alkaloids (monomers). A
high degree of cytotoxic activity was found to reside in the high
molecular weight fraction, which was subsequently passed over a
silicic acid column. However, no crystalline entities were ob-

tained.   Thus, this fraction calls for further purification and separation work.

The low molecular weight fraction, representing the majority of total alkaloids, displayed a lower degree of cytotoxicity. Removal of the major compounds in the early stages of separation often facilitates the later isolation of bases present at lower concentrations.   Since the major alkaloid in this fraction appeared to be vindolinine, as evidenced by TLC and mass spectrometry, the majority of this base was removed by treatment with hydrochloric acid.   The remaining alkaloid mixture was subjected to the gradient pH separation technic into 13 different fractions, five of which were shown to be cytotoxic.   Ajmalicine, pericalline, tetrahydroalstonine, and a small amount of a new alkaloid designated cathahelenine, were isolated from these fractions[4]. None of these were found to be cytotoxic.

Distribution of Catharanthus trichophyllus Alkaloids

With the exception of the incompletely characterized alkaloid cathahelenine, all of the other bases isolated from this plant, i.e. vindolinine (dihydrochloride), ajmalicine, pericalline and tetrahydroalstonine, have been isolated from both *C. lanceus* and *C. roseus*, and their distribution has been noted previously under the discussions pertaining to these two species.

THE PHYTOCHEMISTRY OF CATHARANTHUS OVALIS

No reports of medicinal folklore or pharmacological activity have been published for this newly described *Catharanthus* species.

Alkaloids of Catharanthus ovalis

Only one paper has been published on this new species, in which the isolation and identification of the known alkaloids vindoline, catharanthine, and coronaridine were reported, and the structure of a new base, i.e. cathovaline, was proposed on the basis of spectral data (UV, IR, NMR and mass). An insufficient quantity of cathovaline was isolated to enable these workers to prepare the necessary derivatives to establish its stereochemistry. As indicated previously, it has now been established that cathovaline from *C. ovalis* and cathanneine from *C. lanceus*[41-43] are identical substances and the name cathovaline has been agreed on as representing this alkaloid by Farnsworth and Potier.

Distribution of Catharanthus ovalis Alkaloids

Vindoline, catharanthine and cathovaline (cathanneine) have been found in other *Catharanthus* species and their distribution has been previously discussed. Coronaridine, on the other hand, is a new base to the genus *Catharanthus*, having been isolated previously from *Ervatamia coronaria (Tabernaemontana coronaria)*[117], *E. divaricata*[117], *E. dichotoma*[118], *Tabernaemontana oppositifolia*[117], *T. psychotrifolia*[117], *T. alba*[88], *T. heyneana*[119-121], *T. laurifolia*[122], *T. mucronata*[123], *T. pandacaqui*[124], *T. pachysiphon*[125], *T. brachyantha*[125], *T. contorta*[125], *T. pendiflora*[125], *T. eglandulosa*[125], *Conopharyngia jollyana*[126], *C. durissima*[127], *Gabunia odoratissima*[60], *Stemmadenia donnell-smithii*[88], *S. obovata*[88], *S. tomentosa* var. *palmeri*[88] and *Voacanga africana*[128].

THE PHYTOCHEMISTRY OF CATHARANTHUS LONGIFOLIUS

Folklore and Biological Activities

No medicinal folkloric uses for this species have been found in the literature. Certain extracts and fractions of *C. longifolius* have been reported to elicit an antitumor effect in laboratory animals[49].

Non-Alkaloid Constituents of Catharanthus longifolius

A preliminary investigation of *C. longifolius* roots in our laboratory resulted in the separation of a considerable quantity of waxes, which were shown to be a complex mixture of *n*-alkanes. *n*-nonacosane and *n*-hentriacontane were isolated in that study[129].

Alkaloids of Catharanthus longifolius

Intensive phytochemical investigations have not been initiated as yet on this species due primarily to a difficulty in obtaining investigational material from Madagascar. The plant has been shown, however, to contain alkaloids[49].

REFERENCES

1. G.H. Svoboda, Excerpta Med. Inter. Congr. Ser., 106, 9 (1966).

2. N.R. Farnsworth, R.N. Blomster and J.P. Buckley, J. Pharm. Sci., 56, 23 (1967).

3. H.K. Kim, R.N. Blomster, H.H.S. Fong and N.R. Farnsworth, Econ. Botany, 24, 42 (1969).

4. A.B. Segelman, Isolation, identification and structure elucidation of components from *Catharanthus lanceus* (Apocynaceae) and *Catharanthus trichophyllus* (Apocynaceae), Ph.D. Dissertation, University of Pittsburgh, Pittsburgh, Pa. (1971).

5. W.M. Fylipiw, N.R. Farnsworth, R.N. Blomster, J.P. Buckley
    and D.J. Abraham, Lloydia, 28, 354 (1965).

6. M. Tin-Wa, H.H.S. Fong, R.N. Blomster and N.R. Farnsworth,
    J. Pharm. Sci., 57, 2167 (1968).

7. T.S. Githens, Drug plants of Africa, University of Pennsyl-
    vania Press, Philadelphia, Pa. (1949).

8. R. Pernet, Mem. Inst. Sci. Madagascar (Ser. B), 8, 7 (1957).

9. L. Aldaba and L. Oliveros-Belardo, Rev. Filipina Med. y Farm.,
    29, 259 (1938).

10. N.R. Farnsworth, Lloydia, 24, 105 (1961).

11. N.R. Farnsworth, The phytochemistry and biological activity
    of *Catharanthus lanceus* (Apocynaceae), in: Plants in the
    Development of Modern Medicine, T. Swain (ed.). Harvard
    University Press, Cambridge, pp. 291-298 (1972).

12. G.H. Svoboda, M. Gorman and M.A. Root, Lloydia, 27, 361
    (1964).

13. N.R. Farnsworth, G.H. Svoboda and R.N. Blomster, J. Pharm.
    Sci., 57, 2174 (1968).

14. Anon., Cancer Chemother. Rept., 25, 1 (1962).

15. W.D. Loub, N.R. Farnsworth, R.N. Blomster and W.W. Brown,
    Lloydia, 27, 470 (1964).

16. W.M. Maloney, N.R. Farnsworth, R.N. Blomster, D.J. Abraham,
    and A.G. Sharkey, Jr., J. Pharm. Sci., 54, 1166 (1965).

17. G.H. Svoboda, M. Gorman and R.H. Tust, Lloydia, 27, 203
    (1964).

18. M. Gorman, R.H. Tust, G.H. Svoboda and J. LeMen, Lloydia,
    27, 214 (1964).

19. N.R. Farnsworth, G.H. Svoboda and R.N. Blomster, J. Pharm.
    Sci., 57, 2174 (1968).

20. R.D. Sofia, A.B. Segelman, N.R. Farnsworth and J.P. Buckley,
    J. Pharm. Sci., 60, 1240 (1971).

21. R. Paris and H. Moyse-Mignon, Ann. Pharm. Franc., 14, 464
    (1956).

22. N.A. Pilewski, Studies on *Catharanthus lanceus* root alka-
    loids, M.S. Thesis, University of Pittsburgh, Pittsburgh,
    Pa. (1963).

23. A.B. Segelman, N.R. Farnsworth, H.H.S. Fong and D.J. Abra-
    ham, Lloydia, 33, 25 (1970).

24. R.E. Martello, A phytochemical investigation of *Catharanthus
    lanceus* root alkaloids, M.S. Thesis, University of Pitts-
    burgh, Pittsburgh, Pa. (1964).

25. M.-M. Janot, J. LeMen and Y. Hammouda, Ann. Pharm. Franc.,
    14, 341 (1956).

26. M.-M. Janot and J. LeMen, Compt. Rend., 239, 1311 (1954).

27. R. Paris and H. Moyse, Compt. Rend., 245, 1265 (1957).

28. R. Pernet, G. Meyer, J. Bosser and G. Ratsiandavana, Compt.
    Rend., 243, 1352 (1956).

29. R.N. Blomster, R.E. Martello, N.R. Farnsworth and F.J. Draus,
    Lloydia, 27, 480 (1964).

30. M.-M. Janot, J. LeMen and Y. Gabbai, Ann. Pharm. Franc.,
    15, 474 (1957).

31. D.J. Abraham and N.R. Farnsworth, J. Pharm. Sci., 58, 694
    (1968).

32. N.R. Farnsworth, H.H.S. Fong and R.N. Blomster, Lloydia, 29,
    343 (1966).

33. R.N. Blomster, N.R. Farnsworth and D.J. Abraham, J. Pharm.
    Sci., 56, 284 (1967).

34. D.J. Abraham, N.R. Farnsworth, R.N. Blomster and R.E. Rhodes,
    J. Pharm. Sci., 56, 401 (1967).

35. E.M. Maloney, H.H.S. Fong, N.R. Farnsworth, R.N. Blomster
    and D.J. Abraham, J. Pharm. Sci., 57, 1035 (1968).

36. N.R. Farnsworth, W.D. Loub, R.N. Blomster and M. Gorman,
    J. Pharm. Sci., 53, 1558 (1964).

37. D.J. Abraham, N.R. Farnsworth, R.N. Blomster and A.G. Shar-
    key, Jr., Tetrahedron Lett., 1965, 317.

38. D.J. Abraham, N.R. Farnsworth, W.D. Loub and R.N. Blomster,
    J. Org. Chem., 34, 1575 (1969).

39. R.N. Blomster, N.R. Farnsworth and D.J. Abraham, Naturwissenschaften, 55, 298 (1968).

40. N.R. Farnsworth, W.D. Loub, R.N. Blomster and D.J. Abraham, Z. Naturforsch. 23b, 1061 (1968).

41. G.H. Aynilian, N.R. Farnsworth, R.L. Lyon and H.H.S. Fong, J. Pharm. Sci., 61, 298 (1972).

42. G.H. Aynilian, M. Tin-Wa, N.R. Farnsworth and M. Gorman, Tetrahedron Lett., 1972, 89.

43. G.H. Aynilian, B. Robinson, N.R. Farnsworth and M. Gorman, Tetrahedron Lett., 1972, 391.

44. N. Langlois and P. Potier, Compt. Rend. Ser. C., 273, 994 (1971).

45. G.H. Svoboda, N. Neuss and M. Gorman, J. Amer. Pharm. Assoc., Sci. Ed., 48, 659 (1959).

46. N.R. Farnsworth, R.N. Blomster, D. Damratoski, W.A. Meer and L.V. Cammarato, Lloydia, 27, 302 (1964).

47. R. Munier and M. Macheboeuf, Bull. Soc. Chim. Biol., 33, 846 (1951).

48. G.H. Svoboda, A.T. Oliver and D.R. Bedwell, Lloydia, 26, 141 (1963).

49. N.R. Farnsworth, Unpublished data.

50. N.R. Farnsworth, R.N. Blomster, A.N. Masoud and I. Hassan, Lloydia, 30, 106 (1967).

51. G.H. Svoboda, Lloydia, 24, 173 (1961).

52. G.H. Svoboda, J. Pharm. Sci., 52, 407 (1963).

53. G.H. Svoboda, M. Gorman, N. Neuss and A.J. Barnes, Jr., J. Amer. Pharm. Assoc., Sci. Ed., 50, 409 (1961).

54. M. Gorman, N. Neuss and N.J. Cone, J. Amer. Chem. Soc., 87, 93 (1965).

55. G.H. Svoboda, J. Amer. Pharm. Assoc., Sci. Ed., 47, 834 (1958).

56. M. Tin-Wa, N.R. Farnsworth, H.H.S. Fong and J. Trojánek, Lloydia, 33, 261 (1970).

57. M.-M. Janot, J. LeMen and C. Fan, Bull. Soc. Chim. Franc.,
     1959, 891.

58. M. Gorman, N. Neuss, G.H. Svoboda, A.J. Barnes, Jr. and
     N.J. Cone, J. Amer. Pharm. Assoc., Sci. Ed., 48, 256 (1959).

59. B.K. Moza and J. Trojánek, Chem. Ind. (London), 1962, 1425.

60. M.P. Cava, S.K. Talapatra, J.A. Weisbach, B. Douglas, R.F.
     Raffauf and J.L. Beal, Tetrahedron Lett., 1965, 931.

61. B.K. Moza and J. Trojánek, Collect. Czech. Chem. Commun.,
     28, 1419 (1963).

62. B. Pyuskyuler, I. Kompis, I. Ognyanov and G. Spiteller,
     Collect. Czech. Chem. Commun., 32, 1289 (1967).

63. B.J. Gilbert, J.A. Brissolese, J. Campello, E.P. Carrazzoni,
     R.J. Owellen, E.C. Blossey, K.S. Brown and C. Djerassi,
     Tetrahedron, 21, 1141 (1965).

64. R.R. Arndt and C. Djerassi, Experientia, 21, 566 (1965).

65. U. Renner and R. Keonweisz, Experientia, 19, 244 (1963).

66. A. Walser and C. Djerassi, Helv. Chim. Acta, 48, 391 (1965).

67. M. Shimizu and F. Uchimaru, Chem. Pharm. Bull. (Tokyo), 6,
     324 (1958).

68. G.H. Svoboda, J. Amer. Pharm. Assoc., Sci. Ed., 46, 508
     (1957).

69. J.M. Müller, Experientia, 13, 479 (1957).

70. F.A. Hochstein, J. Amer. Chem. Soc., 77, 5744 (1955).

71. S.C. Pakrashi, C. Djerassi, R. Wasicky and N. Neuss, J. Amer.
     Chem. Soc., 77, 6687 (1955).

72. M.B. Patel, J. Poisson, J. L. Poosset, and J.M. Rowson, J.
     Pharm. Pharmacol., 16, 163 (1964).

73. L. Merlini, R. Mondelli, G. Nasini and M. Hesse, Tetrahedron,
     26, 2259 (1970).

74. M.-M. Janot and J. LeMen, Compt. Rend., 243, 85 (1955).

75. S. Siddiqui and R.H. Siddiqui, J. Indian Chem. Soc., 8, 667
     (1931).

76. H.R. Arthur and S.N. Loo, Phytochemistry, 5, 977 (1966).

77. J.C. do Vale, Garcia Orta, 11, 107 (1963). Through Chem.
     Abstr., 62: 8933.

78. C.-H. Wei, Yao Hsueh Pao, 12, 429 (1965). Through Chem.
     Abstr., 63: 16779.

79. C.W. Losand and W.E. Court, Planta Med., 17, 164 (1969).

80. J. Keck, Naturwissenschaften, 42, 391 (1955).

81. A. Hofmann, Helv. Chim. Acta, 38, 536 (1955).

82. A.S. Belikov, Khim. Prir. Soedin, 5, 64 (1969). Through
     Chem. Abstr., 70: 112375c.

83. R.S. Martin and L. Batlori, Kongr. Pharm. Wiss., Vortr.
     Originalmitt., 23, Muenster (Westfalen), Ger. 1963, 179.
     Through Chem. Abstr. 62: 5140.

84. K. Yamaguchi and H. Shoji, Eisei Shikenjo Hokoku, 1958(76),
     99.

85. F.A. Hochstein, K. Murai and W.H. Boegemann, J. Amer. Chem.
     Soc., 77, 3551 (1955).

86. E.E. Van Tamelen, J. Amer. Chem. Soc., 79, 5256 (1957).

87. D.S. Rao and S.B. Rao, Indian J. Pharm., 18, 202 (1956).

88. O. Collera, F. Walls, A. Sandoval, F. Garcia, J. Herran and
     M.C. Perezamado, Bol. Inst. Quim. (Mexico), 14, 3 (1962).

89. E.J. Shellard, A.H. Beckett, P. Tantivatana, J.D. Phillip-
     son and C.M. Lee, J. Pharm. Pharmacol., 18, 553 (1966).

90. A.H. Beckett, E.J. Shellard, J.D. Phillipson and C.M. Lee,
     Planta. Med., 14, 277 (1966).

91. E.J. Shellard, A.H. Beckett, P. Tantivatana, J.D. Phillip-
     son and C.M. Lee, Planta Med., 15, 245 (1967).

92. T.H. van der Meulen and G.J.M. van der Kerk, Rec. Trav.
     Chim. Pays-bas, 83, 141 (1964).

93. N.J. Dastoor, A.A. Gorman and H. Schmid, Helv. Chim. Acta,
     50, 213 (1967).

94. P.R. Benoin, R.H. Burnell and J.D. Medina, Can. J. Chem., 45, 725 (1967).

95. K.H. Palmer, Can. J. Chem., 42, 1760 (1964).

96. R. Paris and R. Goutarel, Ann. Pharm. Franc., 16, 15 (1958).

97. K. Biemann, M. Friedmann-Spiteller and G. Spiteller, Tetrahedron Lett., 1961, 485.

98. L. Tihon, Bull. Agr. Congo. Belge., 43, 797 (1952). Through Chem. Abstr., 47: 1338d.

99. A. Schomer, Pharm. Zentralhalle, 63, 385 (1922). Through Chem. Abstr., 16: 3526.

100. C.R.C. Boeckel, Semana Med. (Buenos Aires), 120, 316 (1962). Through Chem. Abstr., 57: 2332e.

101. R.F. Bader, D.F. Dickel and E. Schlittler, J. Amer. Chem. Soc., 76, 1695 (1954).

102. R.M. Bernal, A. Villegas-Castillo and O.P. Espejo, Experientia, 16, 353 (1960).

103. N.A. Chaudhury and A. Chatterjee, J. Sci. Ind. Res. (India), 18B, 130 (1959). Through Chem. Abstr., 53: 19295d.

104. E. Haack, A. Popelak and H. Spingler, Naturwissenschaften, 43, 328 (1956).

105. A. Stoll and A. Hofmann, Soc. Biol. Chemists, India, 1955, 248. Through Chem. Abstr., 51: 669z.

106. M. Ishidate, M. Okada and K. Saito, Pharm. Bull. (Japan), 3, 319 (1955).

107. J.S.E. Holker, W.J. Ross, W.B. Whalley and R.F. Raffauf, Phytochemistry, 3, 361 (1964).

108. C. Stellfeld, Tribuna Farm. (Brazil), 30, 35 (1962). Through Chem. Abstr., 58: 12860d.

109. R.F.A. Altman, Inst. Nacl. Pesquisas Amazonia, Publ. No. 1, 1958, 3. Through Chem. Abstr., 52: 17613f.

110. D. Stauffacher, Helv. Chim. Acta, 44, 2006 (1961).

111. Dymock, W.C., J.H. Wooden and D. Hooper, Pharmacographica Indica, Vol. 2, K. Paul, Trench, Trubner and Co., Ltd., London (1891).

112. Chopra, R.N., I.C. Chopra, K.L. Handa and L.D. Kapur, Indigenous Drugs of India, 2nd Ed., U.N. Dhur and Sons, Ltd., Calcutta (1958).

113. A.R. Battersby and R.S. Kapil, Tetrahedron Lett., 1965, 3529.

114. Majumdar, D.N. and B. Paul, Indian J. Pharm. 21, 255 (1959).

115. M. Gabbai, Les Alcaloides des Pervenches Vinca et Catharanthus (Apocynacees), These, Universite de Paris, Faculte de Pharmacie, Serie U. n°291 (1958).

116. A. Klasek, Separation Sci., 3, 319 (1968).

117. M Gorman, N. Neuss, N. J. Cone and J. A. Deyrup, J. Amer. Chem. Soc., 82, 1142 (1960).

118. S. M. Kupchan, A. Bright and E. Macko, J. Pharm. Sci., 52, 598 (1963).

119. N. Ramiah and J. Mohandas, Indian J. Chem., 42, 99 (1966).

120. T.R. Govindachari, B. S. Joshi, A. K. Saksena, S. S. Sathe and N. Viswanathan, Chem. Commun., 1966, 97.

121. E. T. Verkey, P. P. Pillay, A. K. Bose and K. G. Das, Indian J. Chem., 4, 332 (1966).

122. M. P. Cava, S. K. Moudood and J. L. Beal, Chem. Ind. (London), 1965, 2064.

123. A. C. Santos, G. Aguilar-Santos and L. L. Tibayan, Anales Real Acad. Farm., 31, 3 (1965).

124. G. Aguilar-Santos, A. C. Santos and L. M. Jason, J. Philippine Pharm. Assoc., 50, 321 (1964).

125. M. B. Patel, C. Miet and J. Poisson, Ann. Pharm. Franc., 25, 379 (1967).

126. C. Hootele, J. Pecher, R. H. Martin, G. Spiteller and M. Spiteller-Friedmann, Bull. Soc., Chim. Belges., 73, 634 (1964).

127. B. Das, E. Fellion and M. Plat, Compt. Rend., Ser. C.,
      264, 1765 (1967).

128. D. W. Thomas and K. Biemann, Lloydia, 31, 1 (1968).

129. N. R. Farnsworth, F. H. Pettler, H. Wagner, L. Hörhammer
      and H. P. Hörhammer, Phytochemistry, 7, -87 (1968).

130. P. Rasoanaivo, N. Langlois and P. Potier, Phytochemistry,
      11, 2616 (1972).

131. P. Rasoanaivo, N. Langlois and P. Potier, Tetrahedron Lett.,
      1973: 1425.

132. A. Rabaron, M. Plat and P. Potier, Plant. Med. Phytother.,
      7, 53 (1973).

133. N. Langlois and P. Potier, Phytochemistry, 11, 2617 (1972).

134. A. Chatterjee, G. K. Biswas and A. B. Kundu, Indian J. Chem.,
      11, 7 (1973).

ADDENDA

After this chapter was prepared, several papers were
published in which additional alkaloids from *Catharanthus*
species were reported.

Catharanthus longifolius

Potier *et al.* [130-132] isolated the following monomeric
indole alkaloids of diverse skeleta, *viz.* vindolinine, perivine,
catharanthine, normacusine B, pericyclivine, akuammidine,
akuammicine, 2,16-dihydroakuammicine, 2,16-dihydro-N$_a$-methyl
akuammicine (or its isomer), vindorosine, vindoline, pericalline, antirhine, minovincinine, ajmalicine, echitovenine,
vindolicine, desacetylvindoline, cathovaline and 15-hydroxy-
kopsinine.

Catharanthus ovalis

Potier *et al.* [133] described the isolation and characteri-
zation of four additional monomeric alkaloids from this species,
viz. venalstonine, venalstonidine, vindolinine and serpentine,
in addition to the dimeric alkaloids vindolicine, vincaleukoblastine, leurosine and catharine.

Chatterjee *et al.* [134] have reported on the isolation of
rauwolscine, vindoline, demethoxyvindoline, and a new base A.
Base A ($C_{21}H_{26}N_2O_3$), m.p. 269° (decomp.), has a yohimbanoid
structure on the basis of interpretation of ir, uv, mass and
NMR spectroscopic data.

CHAPTER IV

STRUCTURE ELUCIDATION AND CHEMISTRY OF THE BIS CATHARANTHUS ALKALOIDS

Donald J. Abraham

Department of Medicinal Chemistry
University of Pittsburgh
Pittsburgh, Pennsylvania 15213

The bis indole alkaloids from *Catharanthus* species constitute
one of the most interesting areas of phytochemistry because these
molecules are among the most complex nonpolymeric chemical substances
of natural origin, and because they are of biological significance
in the area of cancer chemotherapy (1-8). Neuss (9) *et al.*, and
Svoboda (10) have classified the dimeric *Catharanthus* indoles into
two groups (see Table I); those with an indole-indoline class, and
those miscellaneous bis alkaloids not of the indole-indoline class.
We shall concern ourselves in this report primarily with the indole-
indoline class of bis alkaloids since little chemistry has been
reported on the latter class. For clarity, this chapter is divided
into two parts, namely, structure elucidation and chemistry.

STRUCTURE ELUCIDATION
    One of the first bis alkaloids to be of structural interest
in the *Catharanthus* species (due to its antineoplastic activity)
was leurosine (vinleurosine, VLR). The exact structural assignment
of leurosine is still debated after thirteen years and indicates
the complexity of problems in this area.

TABLE I

Bis Indole-Indoline Alkaloids

	Formula	pK_a	M.P., °C	Source
1. Carosine	$C_{46}H_{56}N_4O_{10}$	4.4, 5.5	214–218	L.
2. Catharicine	$C_{46}H_{52}N_4O_{10}$	5.3, 6.3	231–234 (dec.)	I.
3. Catharine	$C_{46}H_{54}N_4O_{10}$	5.34	271–275 (dec.)	L.
4. Desacetyl VLB (·H_2SO_4)	$C_{44}H_{56}N_4O_8 \cdot H_2SO_4$	5.40, 6.90	>320 (dec.)	L.
5. Isoleurosine	$C_{46}H_{60}N_4O_9$	4.8, 7.3	202–206 (dec.)	L.
6. Leurocristine	$C_{46}H_{56}N_4O_{10}$	5.0, 7.4	218–220 (dec.)	L., R.
7. Leurosidine	_____	5.0, 8.8	208–211 (decl)	L., R.
8. Leurosine	$C_{46}H_{56}N_4O_9 \cdot 8H_2O$	5.5, 7.5	202–205 (dec.)	L., R.
9. Leurosivine (·H_2SO_4)	$C_{41}H_{54}N_3O_9 \cdot H_2SO_4$	4.80, 5.80	>335 (dec.)	R.
10. Neoleurocristine	$C_{46}H_{56}N_4O_{12}$	4.68	188–196 (dec.)	L.
11. Neoleurosidine	$C_{48}H_{62}N_4O_{11}$	5.1	219–225 (dec.)	L.
12. Pleurosine	$C_{46}H_{56}N_4O_{19}$	4.4, 5.55	191–194 (dec.)	L.
13. Rovidine (·H_2SO_4)	_____	4.82, 6.95	>320 (dec.)	L.
14. Vinaphamine	_____	5.15, 7.0	229–235	L.
15. Vincaleukoblastine	$C_{46}H_{58}N_4O_9 \cdot (C_2H_5)_2O$	5.4, 7.4	201–211	L., R.
16. Vincathicine (·H_2SO_4)	_____	5.10, 7.05	>320 (dec.)	L.

Miscellaneous Bis Alkaloids

	Formula	pK_a	M.P., $^{\circ}C$	Source*
1. Carosidine	——	indeterminate	263-278, 283 (dec.)	L., R.
2. Vincamicine	——	4.80, 5.85	224-228 (dec.)	L.
3. Vincarodine	$C_{44}H_{52}N_4O_{10}$	5.8	253-256 (dec.)	L.
4. Vindolicine	$(C_{25}H_{22}N_2O_6)_2$	5.4	248-251 (melts, recryst.) 265-267 (dec.)	L.
5. Vindolidine	$C_{48}H_{64}N_4O_{10}$	4.7, 5.3	244-250 (dec.)	L.
6. Vinosidine	$C_{44}H_{52}N_4O_{10}$ (?)	6.80	253-257	R.
7. Vinsedicine	(Mol. wt. 780)	4.45, 7.35	206	S.
8. Vinsedine	(Mol. wt. 778)	4.65, 7.0	298-200	S.

*S, seeds

The earliest proposals for structural assigments of this
class of alkaloids were made by Neuss, Gorman, Boaz and Cone (11)
for the structures of vincaleukoblastine (VLB) and leurocristine
(vincristine, VCR). These workers concluded that catharanthrine
(I) was the partner of vindoline (II) in the structures of VLB
and VCR.

I   II

III   Early proposal for VLB

The essential evidence was that a common product, cleavamine (III)
was obtained from a similar reaction of leurosine (closely related
to VLB), or I. A close relation between VLB and VCR was also
established, but at that time, evidence for stereochemical assign-
ments was lacking. In the next proposal for the structures of VLB
and VCR (by Neuss *et al.* (12) and Bommer *et al.* (13)) catharanthrine
(I) was replaced by velbanamine (IV) for which steric conformations

IV

at several positions were assigned with good chemical evidence (Fig. 1a). However, the stereochemistry at several positions in the structure of vindoline was revised, but no chemical evidence was given for these changes.

In retrospect, the absolute configuration of III had been determined by X-ray crystallography (14, 15), whereby cleavamine showed one asymmetric center in common with the nine of VLB or VCR and III shared one asymmetric center with the three of velbanamine (IV).

During the same period, Hargrove (16) made an additional proposal (V) for the structure of vindoline, giving reasons for the steric assignment of the OH at position 3. The question of the stereochemistry of these bis alkaloids was finally solved by

X-ray crystallography on the methiodide of leurocristine. Moncrief and Lipscomb (17, 18) then found that the chemical formula and bonds were in agreement with the latest Neuss *et al.* structure, except for the stereochemistry (Fig. 1b). The proposal for two of the three asymmetric centers of velbanamine was correct, but the wrong absolute configuration was chosen. Moncrief and Lipscomb also showed that the chemical proposals for four of the six asymmetric centers of the vindoline portion were correct proposals, without an explicit statement of chemical evidence, but had the wrong absolute configuration. A comparison of the correct structures for VLB and VCR by X-ray can be compared with that obtained by chemical and spectral methods in Figures 1A and B.

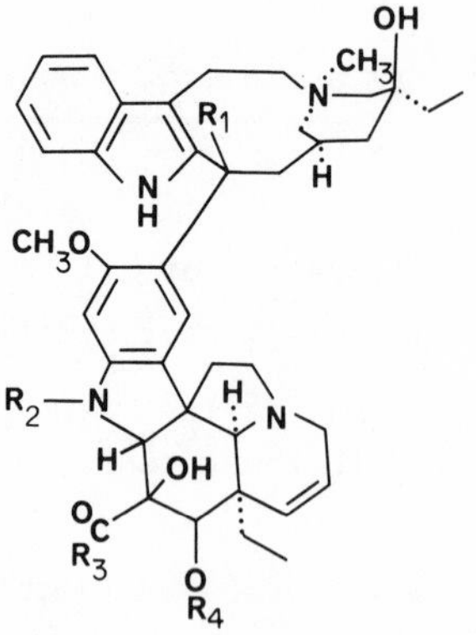

Donald J. Abraham

Fig. 1A. Structure of VCR
methiodide as determined
by X-ray crystallographic
methods. VCR and VLB have
no methyl groups on 6'. The
R groups are defined under
Fig. 1b and the bond linking
15 and 18' has been rotated
by 180° to facilitate
comparision with Fig. 1b.

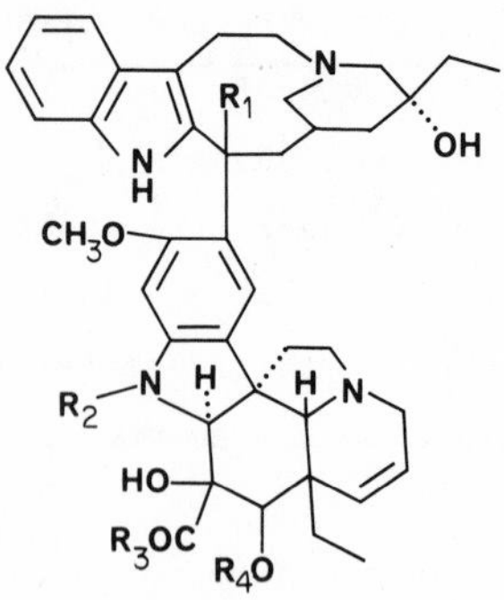

Fig. 1B. Proposal by Neuss
et al. (1964) for the
structures of VLB and VCR.
For VLB: R_1=COOCH$_3$, R_2=
CH$_3$, R_3=OCH$_3$ and R_4=COCH$_3$.
For VCR: R_1=COOCH$_3$, R_2=CHO,
R_3=OCH$_3$ and R_4=COCH$_3$.

After the X-ray analysis established the structures of VLB and
VCR, the structure of leurosidine (vinrosidine, VRD) was suggested
by Neuss, Huckstep and Cone (19), at the 3' OH (VI) isomer of VLB.
The authors state that the hydroxy at C-3' is probably α-oriented,
since a Dreiding model of vinrosamine (VII) with the same relative
stereochemistry of the ethyl group in velbanamine, (VIII), α-
oriented hydroxyl at C-3' presents much more hindrance to approach
for intermolecular hydrogen bonding than in the presence of the
β-oriented hydroxyl.

Of the original four anticancer bis *Catharanthus* alkaloids,
leurosine (VLR) was the last to have a structure assignment.
Leurosine, as the free base, is a very unstable substance which is
decomposed by light, heat, or upon standing in the dark over a
period of a few days. Neuss *et al.* (20) proposed a partial structure

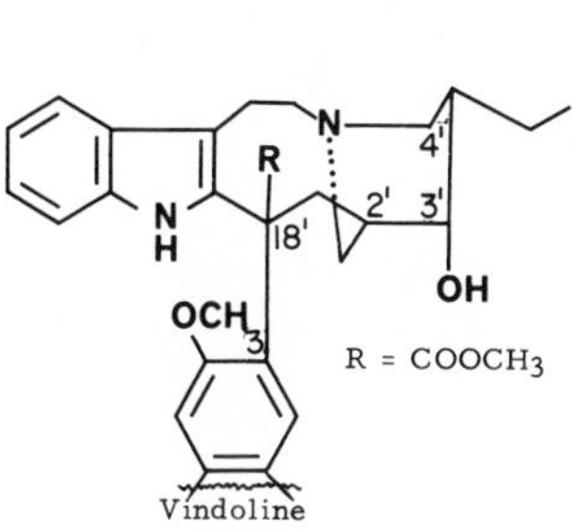

VII Vinrosamine

R = COOCH₃

Vindoline

VI Leurosidine (vinrosidine)   VIII Velbanamine

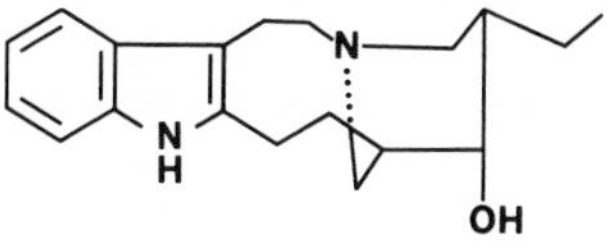

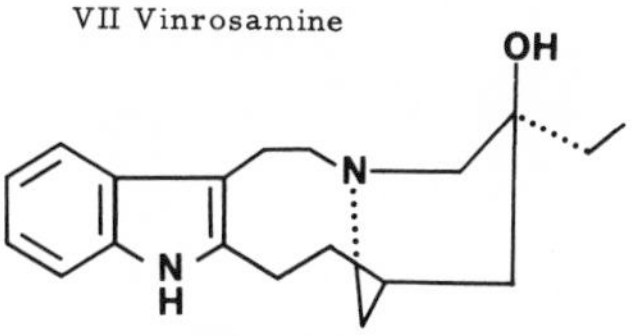

based mainly upon chemical evidence, and Abraham and Farnsworth (21)
proposed a complete structure (IV) which was based primarily on
spectral studies with a detailed account of the mass spectral
fragmentation pattern. The latter represents the most detailed high
resolution mass spectral study performed on a bis *Catharanthus*
alkaloid. Mass spectral fragmentation patterns were correlated for
every major peak in the spectrum and the results, coupled with an
NMR spectrum revealed leurosine to be the epoxy derivative at 3'-4'
position thus correlating nicely with the similar bis alkaloids
VLB (OH at 4') and VRD (OH at 3') alkaloids. Attempts at elucidating

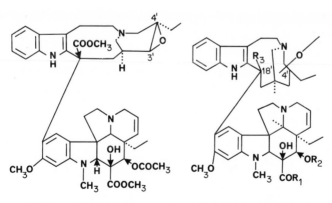

Abraham and Farnsworth   Neuss et al. partial structure
structure for leurosine   for leurosine

the structure of leurosine by X-ray techniques failed in several
laboratories (22-24).

The structure of pleurosine (20) was shown to be the N_b-oxide
of leurosine. Treatment of pleurosine with zinc and acetic acid at
room temperature gave a quantitative yield of leurosine.

The complete structure of catharine has not yet been elucidated, but mass spectral studies by Abraham *et al*. (25) clearly
show that it is made up of one vindoline moiety and an alkaloid
moiety with a molecular formula of $C_{21}H_{23}N_2O_4$.

Isoleurosine was originally thought to be an isomer of leurosine, however, it was found later that isoleurosine was in fact
deoxy VLB (20).

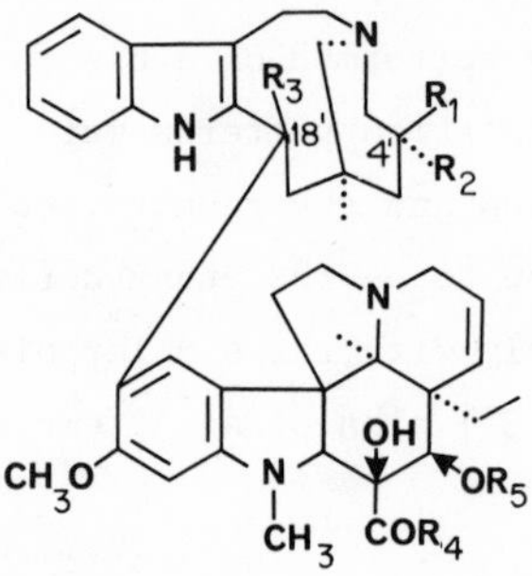

Isoleurosine
(deoxy VLB) "A"

R_1=H, R_2 =C_2H_5,

R_3 = COOCH$_3$

R_4 = OCH$_3$, R_5 = COCH$_3$

MASS SPECTROMETRY

As can be seen by the number of revised structures in this
area of research, the problem of structure elucidation of such
complex molecules (especially when only limited mg. amounts are
available) is difficult. Mass spectrometry has played a key role
in the elucidation of most of the dimerics, and since vindoline
is a partner in most of these dimers, its mass spectral decomposition is summarized below in Scheme I.

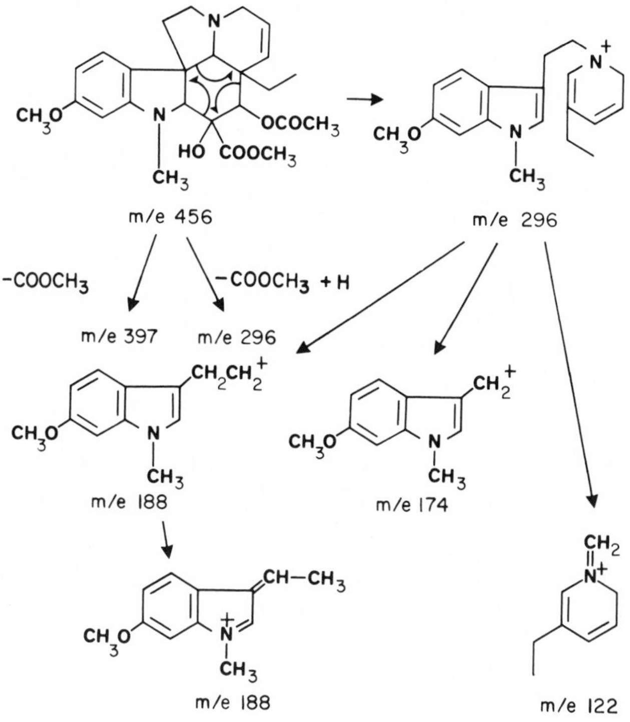

CH3O
N
OCOCH3
HO COOCH3
CH3
m/e 456
CH3O
N
N+
CH3
m/e 296
-COOCH3
-COOCH3 +H
m/e 397
m/e 296
CH2CH2+
CH3O
N
CH3
m/e 188
CH2+
CH3O
N
CH3
m/e 174
CH-CH3
CH3O
N+
CH3
m/e 188
CH2
N+
m/e 122

Mixed ions of vindoline and its partner usually pinpoint functional groups to specific areas thus enabling one to establish a beachhead in the elucidation of the structure. For example, in the structure elucidation of leurosine (21) the following fragmentations could be verified by high resolution measurements.

This fragmentation (Scheme II) is observed since a combination of 156 plus the vindoline ions 455, 396, 395, 295, and 173 from Scheme I give rise to m/e 611, 552, 551, 451, and 329 (Table II).

The combination of m/e 215 with 455, 396, 394, 295, 187, and 173 (Scheme III) was observed to give m/e 670, 611, 610, 510, 402, and 388, all appearing in the spectrum, and the majority of which were measured again by high resolution mass spectrometry for confirmation (Table II).

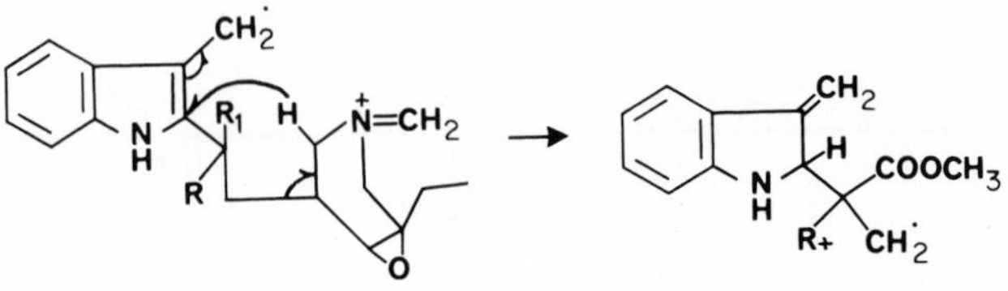

R = vindoline

R_1 = COOCH$_3$

fragment (215) $C_{13}H_{13}NO_2$ + R

R - vindoline ion

Another fragmentation pattern that might arise in this case would be the well-known β-cleavage with γ-hydrogen transfer mechanism, since it is a favored fragmentation route of carbonyl with γ-hydrogens (Scheme IV).

The fragmentation produces an ion at m/e 152 and does agree with the correct molecular formula (Table II).

The positions of eight of the nine oxygens in leurosine have thus been easily discerned from the spectrum. The significant ion m/e 152 then suggests the location of the ninth, or epoxy oxygen. This ion would be similar to the ion m/e 154 obtained from VLB.

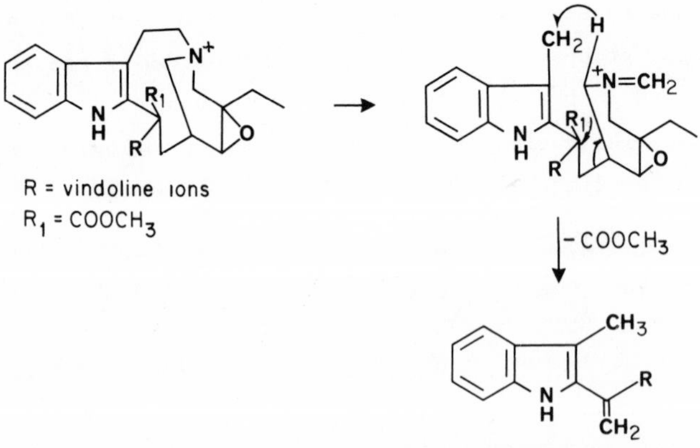

R = vindoline ions

R_1 = COOCH$_3$

- COOCH$_3$

R = m/e vindoline ion
fragment + (156)

TABLE II

High Resolution Measurements on Leurosine

Formula	Observed m/e	Calculated m/e	Formula	Observed m/e	Calculated m/e
C_7H_8N	106.0656	106.0657	$C_{25}H_{28}N_2O_3$	404.2070	404.2098
C_7H_9N	107.0734	107.0735	$C_{30}H_{32}N_3O$	450.2552	450.2543
$C_8H_{10}N$	120.0814	120.0816	$C_{30}H_{33}N_3O$	451.2618	451.2622
$C_8H_{11}N$	121.0894	121.0892	$C_{30}H_{34}N_3O$	452.2714	452.2700
$C_8H_{12}N$	122.0953	122.0970	$C_{29}H_{31}N_3O_3$	469.2342	469.2367
$C_9H_{13}N$	135.1053	135.1048	$C_{32}H_{35}N_3O_3$	509.2673	509.2676
$C_9H_{14}N$	136.1107	136.1126	$C_{32}H_{36}N_3O_3$	510.2747	510.2755
$C_{10}H_{10}N$	144.0814	144.0813	$C_{32}H_{37}N_3O_3$	511.2797	511.2833
$C_9H_{14}NO$	152.1071	152.1075	$C_{32}H_{32}N_3O_4$	522.2427	522.2391
$C_{11}H_8N$	154.0652	154.0656	$C_{33}H_{36}N_3O_3$	522.2759	522.2754
$C_{11}H_{11}N$	157.0880	157.0891	$C_{33}H_{38}N_3O_4$	540.2876	540.2860
$C_{11}H_{12}N$	158.0963	159.0969	$C_{33}H_{39}N_3O_4$	541.2960	541.2938
$C_{11}H_{10}NO$	172.0772	172.0761	$C_{34}H_{36}N_3O_4$	550.2733	550.2704
$C_{12}H_{14}NO$	186.0898	186.0918	$C_{34}H_{37}N_3O_4$	551.2787	551.2782
$C_{22}H_{21}N_2O$	329.1655	329.1653	$C_{34}H_{38}N_3O_4$	552.2852	552.2860
$C_{22}H_{23}N_2O$	331.1814	331.1809	$C_{36}H_{40}N_3O_6$	610.3005	610.2915
$C_{23}H_{24}N_2O$	344.1885	344.1887	$C_{36}H_{41}N_3O_6$	611.3044	611.2993
$C_{23}H_{25}N_2O$	345.1954	345.1965	$C_{36}H_{42}N_3O_6$	612.3149	612.3071
$C_{21}H_{25}N_2O_3$	353.1876	353.1858	$C_{36}H_{43}N_3O_6$	613.3214	613.3149
$C_{25}H_{23}N_2O_2$	383.1778	383.1759	$C_{38}H_{44}N_3O_8$	670.3077	670.3121
$C_{25}H_{25}N_2O_3$	401.1877	401.1864	$C_{38}H_{45}N_3O_4$	671.3170	671.3204
$C_{25}H_{26}N_2O_3$	402.1914	402.1941	$C_{46}H_{56}N_4O_9$	808.4117	808.4044
$C_{25}H_{27}N_2O_3$	403.2049	403.2020			

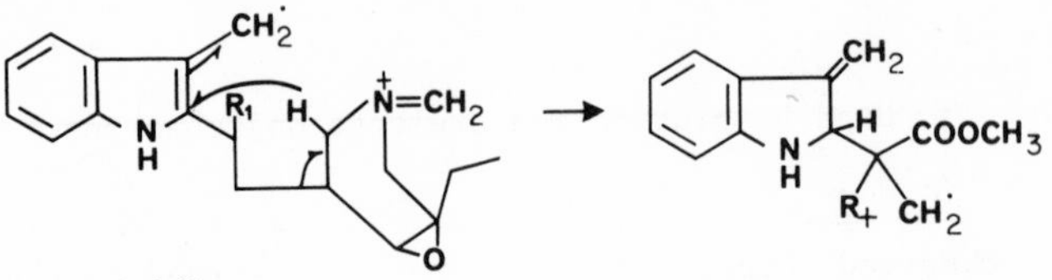

R = vindoline
R_1 = COOCH$_3$

fragment (215) $C_{13}H_{13}NO_2$ + R
R = vindoline ion

The NMR spectrum then gave further evidence for the epoxy group
being attached to the 3'-4' position. The NMR spectrum of leurosine
is very similar to that of VLB. An observable difference is that
of a peak centered around 6.9 τ. This peak was shown to be a doublet
with J equal to 4.1 ± 0.2 c.p.s., both at 60 and 100 MC. The
observed position for an epoxy methine proton of similar type is
reported to be at 6.9-7.2 τ with a J_{cis} of 3.3-4.1 c.p.s. If one
looks at the hydroxyl group in leurosine, one can reasonably assume
that if the hydroxyl of velbanamine was involved in this epoxy
linkage, the methine hydrogen of the epoxy would be *cis* to the
bridge proton, and the J_{cis} coupling constant and chemical shifts

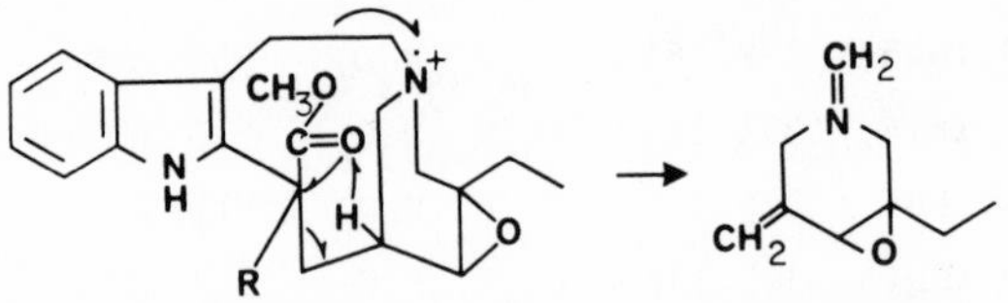

R = vindoline

Ion m/e 152
$C_9H_{14}NO$

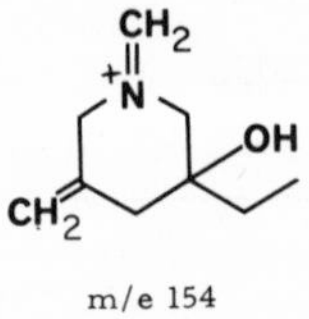

m/e 154

would agree with the proposed structure. This evidence also would
eliminate the other possible attachments of oxygen to any place in
the epoxyvelbanamine structure since either the chemical shifts,
multiplicities, and/or coupling constants would be different. The
integral of this epoxy proton also proves to be equivalent to one
proton when matched against the six methoxyl protons at 6.2 τ.

It is this kind of mass spectral information, coupled with
other techniques, that makes it a realistic endeavor to attempt
the elucidation of such complex structures with only small amounts
of material. Much credit is due to Professors Biemann and Djerassi
and their colleagues for laying the mass spectral ground work in
the elucidation of the structures of alkaloids.

CHEMISTRY OF THE DIMERIC CATHARANTHUS ALKALOIDS

Although much progress has been made in the synthetic area
with the monomeric indole alkaloids, little has been reported with
the dimeric *Catharanthus* alkaloids. Harley-Mason and Rohman (26)
have synthesized demethoxycarbonyldeoxyvinblastine (X) from IX and
vindoline.

Hargrove (27) made a series of acyl and α-aminoacetyl analogs
of VLB and many showed greater activity against the P-1534 mouse
leukemia than that of the parent alkaloid VLB.

Most of the chemistry in this area has resulted from cleavage
reactions of the dimerics to monomeric units. For example, the

following reaction of VLB produces a novel amino acid (XI) and
desacetylvindoline. (12, 28, 29)

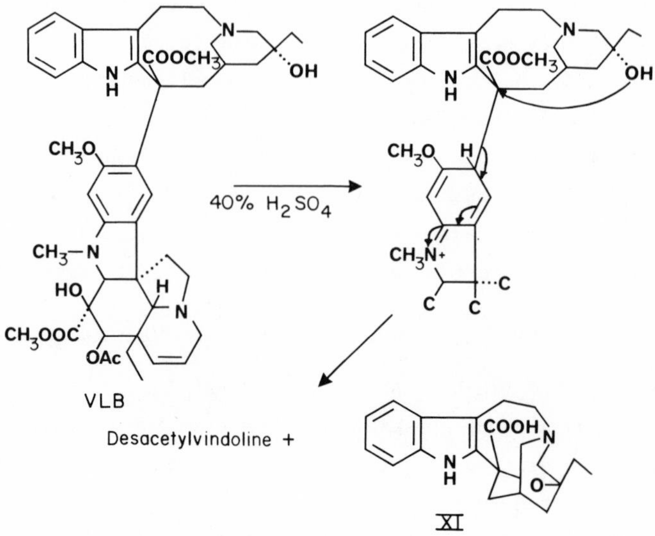

VLB

Desacetylvindoline +

XI

Investigations on further partial syntheses of VLB have not
yet been completed, however, Büchi (28) has outlined a possible
approach.

There has been little activity in the literature on the dimeri‹
Catharanthus alkaloids (chemistry or structure elucidation) during
the last few years. This, however, should not be misinterpreted
to mean that little activity is taking place. The chemical
complexity and biological significance of this class of alkaloids
are of sufficient quality to attract much laboratory activity.
It may, however, take some time until a complete synthesis
emerges, and until complete biogenetic relationships can be
established.

REFERENCES

1. Cutts, J. H., Proc. Am. Assoc. Cancer Res., 2, 289 (1958).

2. Cutts, J. H., Beer, C. T., and Noble, R. L., Rev., Can. Biol., 16, 487 (1957).

3. Cutts, J. H., Beer, C. T., and Noble, R. L., Cancer Res., 20, 1023 (1960).

4. Noble, R. L., Beer, C. T., and Cutts, J. H., Biochem. Pharmacol., 1, 347 (1958).

5. Noble, R. L., Beer, C. T., and Cutts, J. H., Ann. N. Y. Acad. Sci., 76, 882 (1958).

6. Svoboda, G. H., Neuss, N., and Gorman, M., J. Pharm. Sci., 48, 659 (1959).

7. Johnson, I. S., Baker, L. A., and Wright, H. F., Ann. N. Y. Acad. Sci., 76, 861 (1958).

8. Svoboda, G. H., J. Pharm. Sci., 47, 834 (1958).

9. Neuss, N., Johnson, I. S., Armstrong, J. G., and Jansen, C. J., Advances in Chemotherapy, Academic Press, N. Y., N. Y., (1969), p. 133.

10. Svoboda, G. H., Medica, International Congress, Series No. 106, Proceedings First Symposium European Cancer Chemotherapy Group, Paris, June, 1965.

11. Neuss, N., Gorman, M., Boaz, H. E., and Cone, N. J., J. Am. Chem. Soc., 84, 1509 (1962).

12. Neuss, N., Gorman, M., Hargrove, W., Cone, N. J., Biemann, K., Büchi, G., and Manning, R. E., ibid., 86, 1440 (1964).

13. Bommer, P., McMurray, W., and Biemann, K., ibid., 86, 1439 (1964).

14. Kutney, J. P., Trotter, J., Tabata, T., Kerigan, A., and Camerman, N., Chem. and Ind. (London) 648 (1963).

15. Camerman, N., and Trotter, J., Acta Cryst., 17, 384 (1964).

16. Hargrove, W. H., Lloydia, 27, 340 (1964).

17. Moncrief, J. W. and Lipscomb, W. N., J. Am. Chem. Soc.,
     87, 4963 (1965).

18. Moncrief, J. W. and Lipscomb, W. N., Acta Cryst., 21, 322
     (1966).

19. Neuss, N., Huckstep, L. L., and Cone, N. J., Tetrahedron
     Letters, 1967, 811.

20. Neuss, N., Gorman, M., Cone, N. J. and Huckstep, L.L.,
     Tetrahedron Letters, 1968 783.

21. Abraham, D. J. and Farnsworth, N. R., J. Pharm. Sci., 58,
     694 (1969).

22. Abraham, D.J., University of Pittsburgh, personal communi-
     cation.

23. Moncrief, J. W., Emory University, personal communication.

24. Lipscomb, W. N., Harvard University, personal communica-
     tion.

25. Abraham, D.J., Farnsworth, N. R., Blomster, R. N., and
     Rhodes, E. E., J. Pharm. Sci. 56, 401 (1967).

26. Harley-Mason, J., and Rahman, Atta-ur-, J. Chem. Soc. D.
     1967, 1048.

27. Hargrove, W. W., Lloydia, 27, 340 (1964).

28. Büchi, G., Pure Appl. Chem. 9, 21 (1964).

29. Neuss, N., Gorman, M., and Cone, N. J. Lloydia, 27, 389
     (1964).

CHAPTER V

THE BIOSYNTHESIS OF CATHARANTHUS ALKALOIDS

Ronald J. Parry

Department of Chemistry
Stanford University
Stanford, California 94305

INTRODUCTION

An immense variety of secondary metabolites is produced
by the flowering plants. Within this diverse collection of
substances, the indole alkaloids constitute one of the largest
groups as their total number now exceeds 800.[1,2] Interest in
the chemistry of these alkaloids is as old as organic chemistry
itself, but it is only comparatively recently, with the ad-
vent of radiotracer techniques, that information has become
available on the pathways utilized by the Angiosperms in the
elaboration of these complex natural products. The historical
background and earliest investigations in this field having
been carefully summarized,[3,4] this account surveys the recent
progress which has been made in our understanding of *Catharanthus* alkaloid biosynthesis. The literature has been screened
through May, 1971.

ORIGIN OF THE "C_9-C_{10}" UNIT OF THE CATHARANTHUS ALKALOIDS

The alkaloids of *Catharanthus* species and nearly all of
the indole alkaloids occurring in Nature are formally derived

*Present address: Department of Chemistry, Brandeis University,
Waltham, Massachusetts

from the combination of a tryptamine unit with an ubiquitous "C_9-C_{10}" unit. This unit appears in three skeletal guises[5] which are referred to[6] as the *Corynanthe-Strychnos* unit (1), the *Aspidosperma* unit (2), and the Iboga unit (3), the names being derived from the plant genus in which alkaloids of the specific structural type are the most conspicuous members. The three forms of the non-tryptophan derived unit are shown in Scheme I. The dotted line in (1) - (3) indicates that carbon atom which is consistently absent from those alkaloids having only nine skeletal carbons in addition to the tryptamine moiety.

Essentially all the information that is available on the biosynthesis of the indole alkaloids found in *Catharanthus* species has been derived from experimentation with one species, *Catharanthus roseus* G. Don (*Vinca rosea* L.) (Apocynaceae). The concentration of effort on this particular species is undoubtedly the result of three factors: 1) the plant is readily available, 2) it possesses excellent hydroponic qualities, and 3) it produces a broad spectrum of indole alkaloids, over 70 bases having been isolated to date.[1,2] Alkaloids found in *C. roseus* which illustrate each of the three skeletal forms of the "C_9-C_{10}" unit are (Scheme I): ajmalicine (4a) and serpentine (4b) (*Corynanthe*); akuammicine (5) (*Strychnos*); vindoline (6) (*Aspidosperma*) and catharanthine (7) (Iboga); the skeleton in question is emphasized in each structure with heavy bonds.

Early experimental work had demonstrated the ability of the amino acid tryptophan to serve as a precursor of the tryptamine moiety of the indole alkaloids,[7,13] and the principal stumbling block which for some time impeded progress in the field was the origin of the "C_9-C_{10}" unit. The absence of any experimental evidence allowed the formulation of a number of ingenious hypotheses. The earliest proposal[8] suggested that the *Corynanthe-Strychnos* unit originates from 3, 4-dihydroxyphenylalanine (DOPA) plus two "C_1" units *via* a cleavage of the

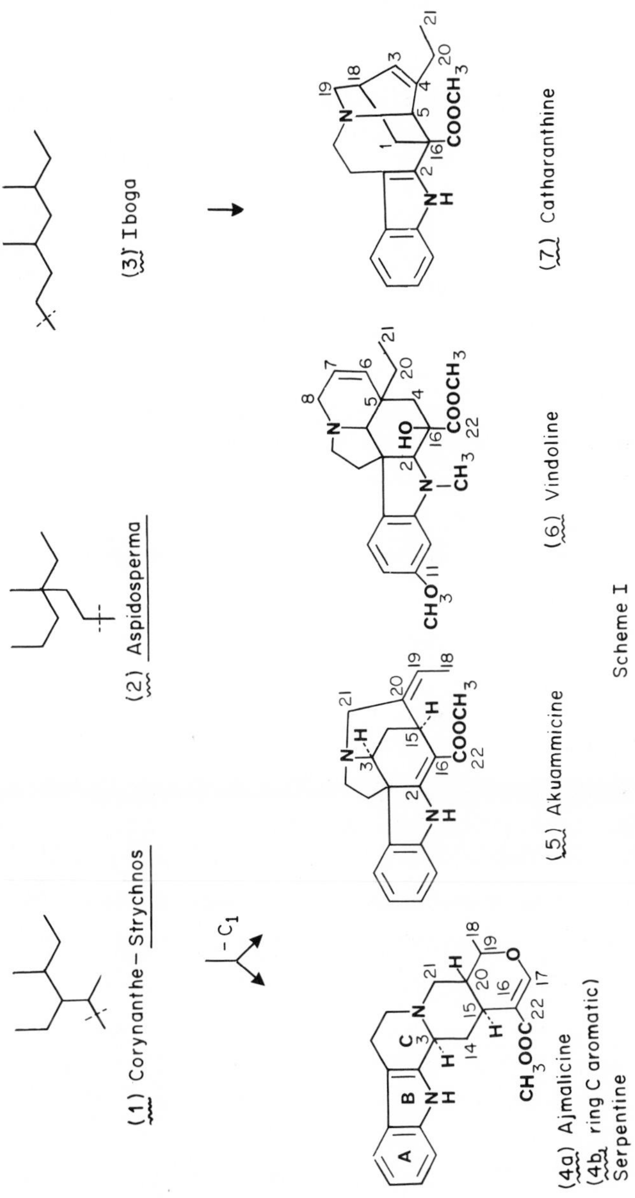

(1) Corynanthe-Strychnos

(2) Aspidosperma

(3) Iboga

(4a) Ajmalicine
(4b) ring C aromatic)
Serpentine

(5) Akuammicine

(6) Vindoline

(7) Catharanthine

Scheme I

aromatic ring between the vicinal hydroxyl groups ("Woodward fission"). A second suggestion[9] derived the required ten carbon skeleton from acetate, malonate, and a "C_1" unit, while a third hypothesis invoked prephenic acid[10] and a one carbon fragment as the likely precursors. Each of these theories has now been disproven.

The acetate hypothesis initially appeared to receive experimental support[9,11,12] when it was reported that the feeding of sodium ^{14}C-formate, sodium 1-^{14}C-acetate, and 1, 3-^{14}C-malonic acid to *Rauvolfia serpentina* Benth. (Apocynaceae) led to specific incorporations into the *Rauvolfia* indole alkaloids. These results could not be reproduced,[13] however, and investigations carried out in other laboratories[14,15] found only scattering of the label when the above precursors were supplied to *Rauvolfia* plants.

The prephenic acid hypothesis was likewise shown to be untenable when it was demonstrated[16] that administration of U-^{14}C-shikimic acid to *Catharanthus roseus* shoots gave radioactive vindoline (6) bearing over 90% of its activity in the benzene ring. The result is in accord with the expected conversion of shikimic acid to aromatic amino-acids *via* prephenic acid,[17] but clearly provides no grounds for assigning prephenic acid the role of precursor of the "C_9-C_{10}" unit.

The genesis of the "C_9-C_{10}" skeleton *via* Woodward fission of DOPA was removed from the realm of possibility as a result of feeding experiments[18] carried out to examine the biosynthesis of cephaeline (8) and emetine (9), the major alkaloids of *Cephaelis ipecacuanha* Rich. (Rubiaceae). These two alkaloids each contain the *Corynanthe* unit (heavy bonds) sequestered by a dopamine residue rather than by tryptamine. The implication is that conclusions derived from tracer studies on (8) and (9) may be applied, with caution, to the analogous *Catharanthus* alkaloids. Specifically, it was found that administration of

(8, R = H) Cephaeline
(9, R = CH) Emetine

$2\text{-}^{14}C$-DL-tyrosine (side-chain label) to *C. ipecacuanha* shoots
yielded radioactive cephaeline which was labeled only at C-1
and C-1', with no activity residing in the carbons of the C_9
skeleton. Similar results were obtained when ^{14}C-labeled phe-
nylalanine was infused into ipecac plants. A scrambling of
label, analogous to that observed in *R. serpentina (vide supra)*,
was the consequence of feeding sodium ^{14}C-formate, sodium 1-
^{14}C-acetate, and sodium 1, $3\text{-}^{14}C$-malonate to *C. ipecacuanha*.

The experimental findings outlined above allowed three
reasonable pathways to the "$C_9\text{-}C_{10}$" unit of the *Catharanthus*
alkaloids to be dismissed. One hypothesis remained to be tested.
This remarkably perceptive suggestion, adumbrated independently
by Thomas[19] and by Wenkert,[10b] derived the mysterious non-tryptophan based carbons by fission of the five-membered ring of a
cyclopentanoid monoterpene. The postulated fission leads to
the *Corynanthe-Strychnos* skeleton and rearrangements were invoked to generate the *Aspidosperma* and Iboga skeletons from it
(Scheme II). On the basis of these speculations, mevalonic
acid (10) was expected to serve as the fundamental precursor
of the non-tryptophan derived carbon atoms of the indole alkaloid skeletons. This expectation has been thoroughly vindicated.

The initial experiments, carried out by administration of
sodium $2\text{-}^{14}C$-mevalonate to *C. roseus* shoots, afforded[20,21,22]
radioactive ajmalicine (4), vindoline (6), catharanthine (7),
and reserpinine (11). Each of these alkaloids was degraded to

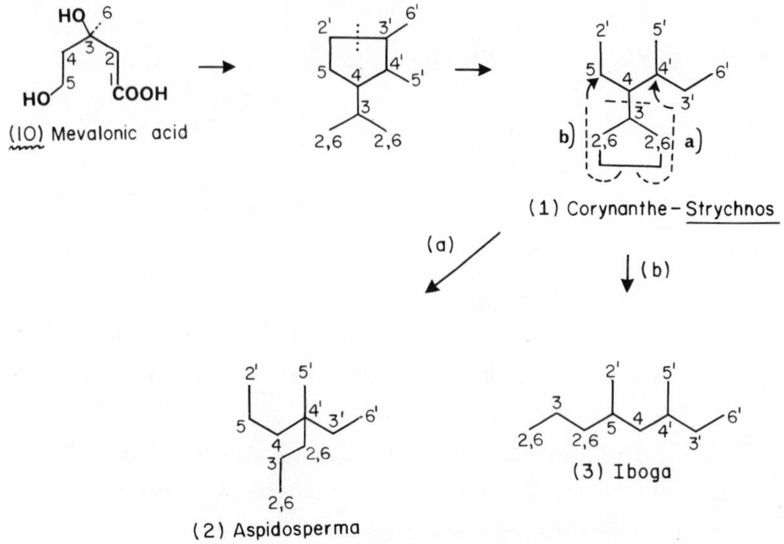

(10) Mevalonic acid

(1) Corynanthe-Strychnos

(2) Aspidosperma

(3) Iboga

Scheme II

show that *ca.* 25% of the respective total activities resided
at C-22. These results indicate that specific incorporation
of mevalonic acid into each of the alkaloids takes place, but
that the biosynthetic processes lead to equilibration of C-2 and
C-6 of one mevalonate unit. Similar behavior had been previously encountered[23] in the biosynthesis of plumeride (12). The
radioactive catharanthine obtained in these feedings was further
degraded to show that 2% of the total activity was at C-1 and/

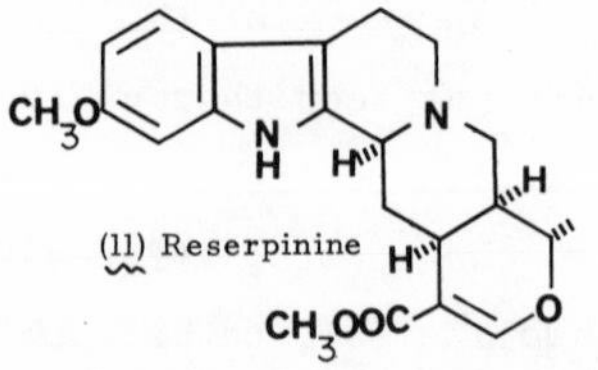

(11) Reserpinine

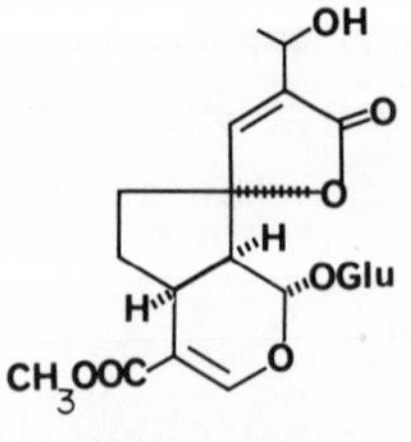

(12) Plumeride

or C-16 (theory: 25% at C-1), and 48% was at one or more of the
carbons, -18, -3, -5, and -19 (expect 50% at C-19).   These labeling patterns are in complete agreement with the formalized
pathways of Scheme II.   This scheme correctly predicted the
distribution of labels resulting from an extensive series[24,25,26]
of ^{14}C-mevalonate feedings to *C. roseus*.   These experimental
findings, which provide a nearly complete mapping of the "C_9-C_{10}"
unit, are summarized in Table I.   In addition, 5-^{2}H$_2$-mevalonate
was supplied to *C. roseus* and the resulting deuterated vindoline (13) analyzed by mass spectrometry.[27]   The principal fragment ions of vindoline are shown in Scheme III.   Ions (a), (b),
and (c) showed a deuterium enrichment corresponding to the presence of two deuterium atoms while ions (d) and (e) showed enrichment.

VERIFICATION OF THE THOMAS-WENKERT HYPOTHESIS

With the role of mevalonic acid as a precursor of the indole alkaloids established beyond question, the suspicion was
aroused that geraniol (15) might serve as the acyclic monoterpene bridge between (10) and a cyclopentanoid monoterpene.
Three groups simultaneously confirmed this suspicion.   2-^{14}C-
Geraniol was administered[24,25,27] to *C. roseus* and found to
yield radioactive ajmalicine (4a), catharanthine (7), and vindoline (6).   The vindoline carried all of its activity at C-5,
the catharanthine was labeled exclusively at C-4, and the
ajmalicine produced was labeled at one or more of four carbons:
-3, -14, -20, and/or -21 (expect C-20 labeling).   Mass spectrometry was again employed to localize the deuterium residing
in the vindoline (14) resulting from the feeding of 1-^{2}H$_2$-ge-
raniol to *C. roseus*;[27] ions (a), (b), and (c) showed, in this
instance, an enrichment in deuterium corresponding to the presence of one atom of the hydrogen isotope, while ions (d) and
(e) showed no such enrichment.   Additional support for the

TABLE I

Administration of ^{14}C-Labeled Mevalonic Acids to *Catharanthus roseus*

Position of ^{14}C Label	Alkaloid Isolated	Carbon Atom(s) Isolated	%Total Activity Theory	%Total Activity Found
2	Ajmalicine (4a)	16,17	25	22
2	Serpentine (4b)	3	50	43
3	Ajmalicine (4a)	19	50	40
3	Serpentine (4b)	19	50	40
3	Catharanthine (7)	20,21	50	44
3	Vindoline (6)	20	50	45
		20	50	47
		5	0	0
		21	0	0
		22	0	0
		0-Methyl	0	0
		N-Methyl	0	0
4	Serpentine (4b)	15	50	45
4	Catharanthine (7)	4	50	48
4	Perivine (27)	20	50	44
5	Serpentine (4b)	14	50	43
5	Catharanthine (7)	4	0	0
		20	0	0
		21	0	0
5	Vindoline (6)	5	0	0
		20	0	0
		21	0	0

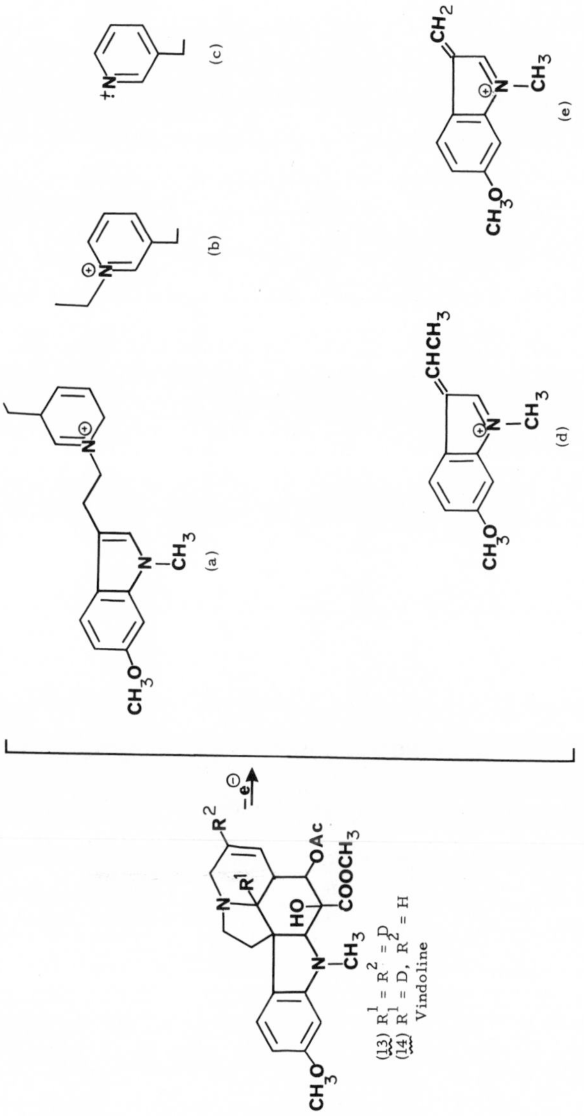

Scheme III

specific incorporation of geraniol into the "C_9-C_{10}" skeleton
was subsequently provided by the report[28] that 3-^{14}C-geraniol,
when administered to *C. roseus*, leads to radioactive catharanthine and vindoline, each labeled only at C-20. All of these
observations are in harmony with predictions made on the basis
of the Thomas-Wenkert hypothesis (see Scheme IV). The incorporation levels of the labeled geraniols are summarized in
Table II.

Final vindication of the Thomas-Wenkert hypothesis required
that a cyclopentanoid monoterpene be found which under went the
postulated ring-fission to generate the *Corynanthe-Strychnos*
unit. A clue to the nature of such a monoterpene was uncovered
when the structure of a nitrogenous glucoside found in *C.
ipecacuanha* was elucidated. This glucoside, called "ipecoside",

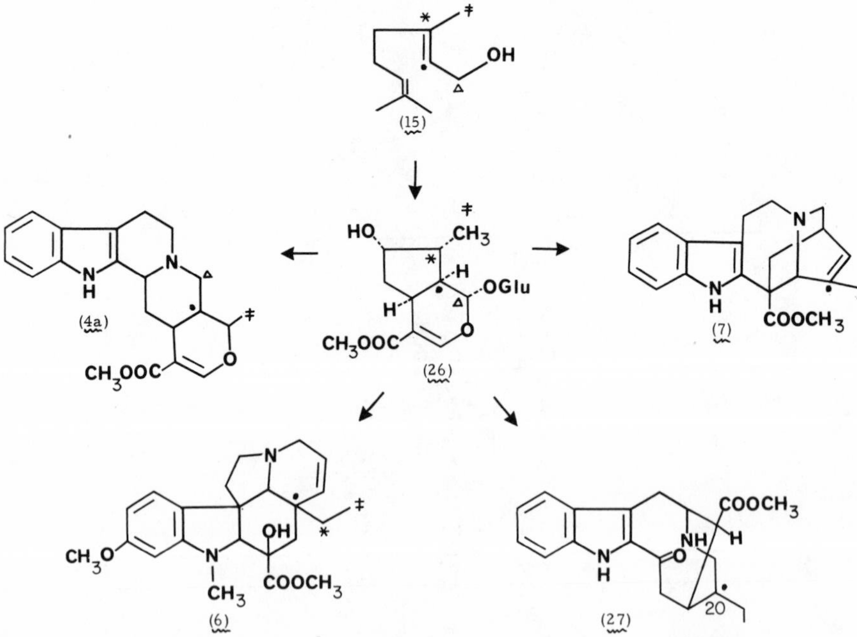

Scheme IV

TABLE II

Incorporations of Geraniol, Loganin, Secologanin, and Vincoside into *Catharanthus* Alkaloids

Precursor	Alkaloid Incorporations (%)				
	Ajmalicine (4a)	Serpentine (4b)	Catharanthine (7)	Vindoline (6)	Perivine (27)
2-^{14}C-Geraniol	0.16	0.60	0.20	0.20 / 0.10	—
1-^{2}H$_2$-Geraniol	—	—	—	0.80	—
O-Me-^{3}H-Loganin	0.026	0.45	0.80	0.50	0.10
2-^{14}C-Loganin	0.46	0.60	0.50	0.23	0.04
4-^{14}C-Loganin	0.10	—	0.30	0.20	—
O-Me-^{3}H-Secologanin	0.55	0.65	0.16	0.12	0.013
O-Me-^{3}H-Vincoside and C-3 Epimer	0.47	3.91	0.84	0.57	0.056
Ar-^{3}H, O-Me-^{3}H-Vincoside and C-3 Epimer	0.65	1.28	0.41	0.31	0.037
O-Me-^{3}H/Ar-^{3}H = 1.44	1.41	1.43	1.45	1.48	1.46

was assigned structure (16, R = α-H) on the basis of chemical
and spectroscopic evidence.[29]   Its stereochemistry was defined
by correlation[30] with protoemetine (17) *via* dihydroprotoemetine
(18) as shown (Scheme V). Very recently, ipecoside has been the
subject of X-ray analysis.[31] This analysis confirmed the gross
structure, but surprisingly, it revealed that the glucoside has
its C-3 hydrogen in the β-configuration rather than in the α-
configuration as suggested by the correlation with dihydroprotoemetine. Ipecoside is therefore correctly represented by (16,
R = β-H). Since the configuration at C-3 of dihydroprotoemetine
has been unequivocally determined by degradation of emetine,[32]
it appears that epimerization at C-3 of ipecoside occurs during
the correlation sequence outlined in Scheme V. The most likely
was in which this may happen is probably *via* the reversible
ring-opening of the intermediate iminium salt (19) to (20).

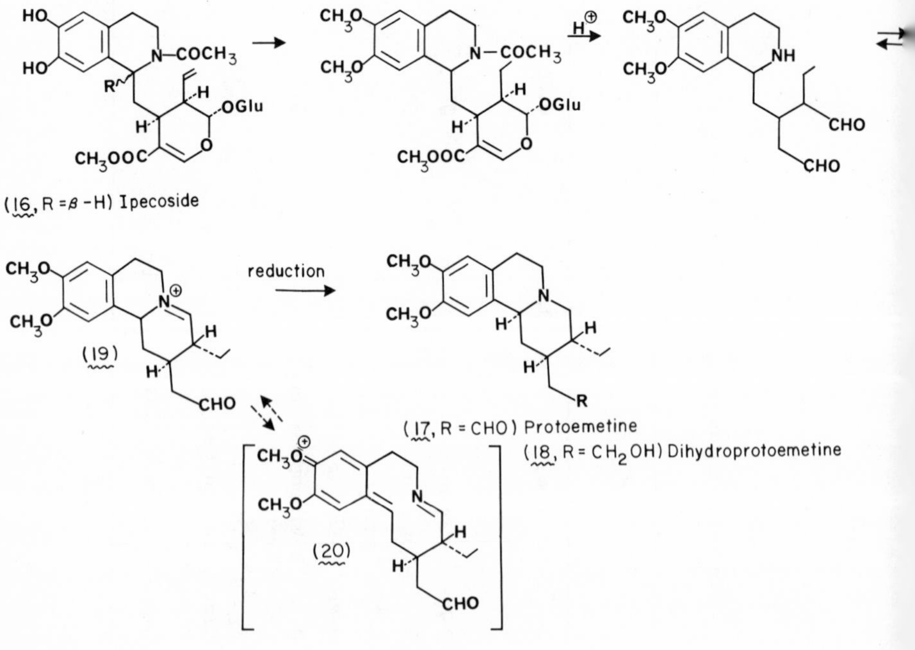

Scheme V

At the time when the structure of ipecoside was first determined, this complication had not yet arisen, and the successful conversion of (16) into (18) as well as the obvious structural similarity between protoemetine and corynantheine (21) suggested that both (17) and (21) could arise in Nature as a consequence of the interaction of the unknown aldehyde (22) with the appropriate amine. The question then arose as to which cyclopentanoid monoterpene could lead to (22) by rupture of the cyclopentane ring. Four iridoid[33] glucosides were considered to be likely candidates.[6] These were verbenalin (23),[34] genepin (24),[35] monotropeine (shown as its methyl ester, 25),[36] and

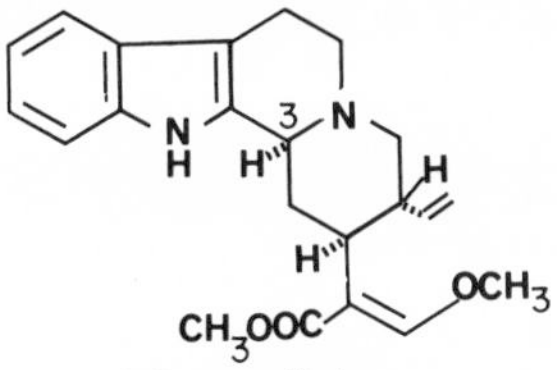

(21) Corynantheine

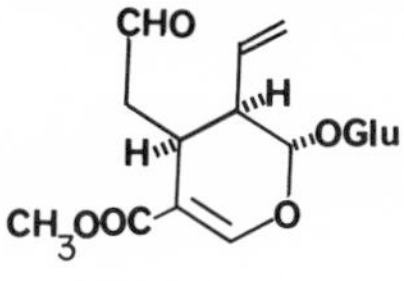

(22) Secologanin

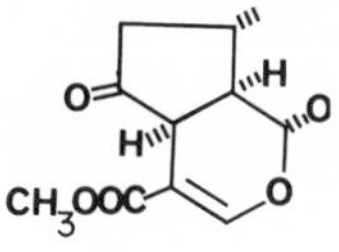

(23) Verbenalin

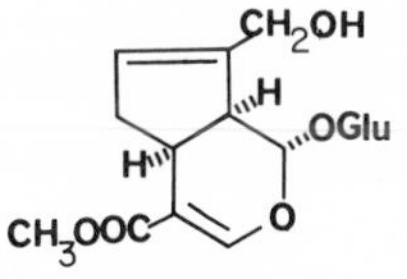

(24) Genepin

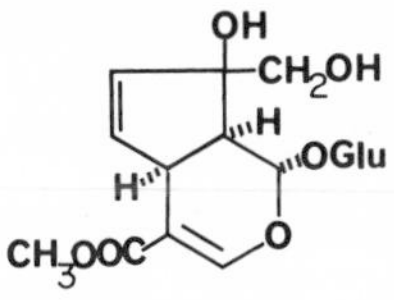

(25) Monotropeine methyl ester

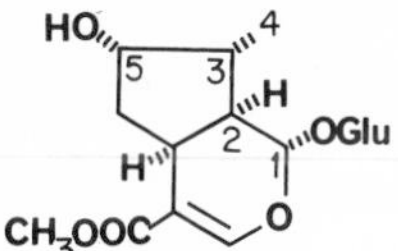

(26) Loganin

loganin (26).[37] Each of these glucosides was labeled with tritium and tested for its ability to serve as the precursor of the indole alkaloids of *C. roseus*. No incorporations resulted when labeled verbenalin, dihydroverbenalin, genepin, or monotropeine methyl ester was supplied to the plant.[6,38] However, when O-methyl-^{3}H-loganin, prepared by hydrolysis of loganin to loganic acid and reesterification with ^{3}H-diazomethane, was administered to *C. roseus*, radioactive ajmalicine (4a), serpentine (4b), perivine (27), vindoline (6), and catharanthine (7) were produced.[38] Zeisel demethylation of (4a), (4b), and (7) demonstrated that all of the radioactivity resided in the methyl group of the ester function present in each alkaloid (Scheme IV). Hydrolysis of the vindoline to desacetylvindoline (carrying 98% of the original activity) followed by reduction with lithium aluminium hydride gave a triol which contained less than 0.1% of the original activity. Specific incorporation of loganin into the *Corynanthe*, *Aspidosperma*, and Iboga families of indole alkaloids was thereby established. The levels of incorporation are listed in Table II.

As further evidence for the role of loganin in indole alkaloid biosynthesis, its presence in *Catharanthus roseus* was proven by isotopic dilution techniques. 1-^{3}H-geraniol was infused into the plants and radio-inactive loganin added to the plant workup. Reisolation of the loganin as its penta-acetate, hydrolysis, and methylation of the loganic acid afforded loganin with a constant specific activity equivalent to 0.02% incorporation of 1-^{3}H-geraniol.[38]

The discovery of the role played by loganin in indole alkaloid biosynthesis prompted a reexamination of its structure. The original structural arguments[34,37] were founded upon the application of biogenetic theory to limited chemical data and left the stereochemistry unassigned. Three laboratories therefore chose to reexamine the problem[39,40,41] and, happily, they each agreed that the original structural assignment was correct as well as arriving at the absolute stereochemistry depicted in (26).

The expected biosynthetic derivation of loganin from geraniol has been established by feeding experiments with the bog-bean, *Menyanthes trifoliata* L. (Gentianaceae), a plant rich in loganin.

When administered to the rhizomes of *Menyanthes*, $4-^{14}$C-geraniol yielded loganin carrying 85% of its label at C-4;[40] rather surprisingly, sodium $2-^{14}$C-mevalonate was not incorporated into loganin under similar conditions, a result best attributed to a failure of the mevalonate to reach the site of loganin biosynthesis. The feeding of a mixture (*ca.* 3:1) of $2-^{14}$C-geraniol and $2-^{14}$C-nerol (isomeric at the 2,3 double-bond) led, as anticipated, to $2-^{14}$C-loganin (see Scheme IV).[39]

The two biosynthetically labeled specimens of loganin were then supplied to *C. roseus* shoots[42,43] and the labeling patterns in the isolated alkaloids found by degradation to agree with the predictions of the Thomas-Wenkert hypothesis (Scheme IV and Table II). Further, *Cephaelis ipecacuanha* plants were shown to convert loganin efficiently[44] (1.7 - 1.9 %) into ipecoside, a result in agreement with the postulated fission of the cyclopentane ring of loganin to generate the aldehyde (22).

Condensation of the aldehyde (22) with tryptamine would be expected to yield the indole analogs of ipecoside, the β-carboline glucosides, (28a) and (28b). The natural occurrence of the glucosides (22) and (28a, 28b) has now been demonstrated. The key to the solution of this problem was discovered when the structures of three glucosides which occur in *Menyanthes trifoliata* were solved.[45,46] These glucosides, foliamenthin (29a), dihydrofoliamenthin (29b), and menthiafolin (30), on careful alkaline hydrolysis followed by diazomethane treatment yield the aldehyde (22), called "seco-loganin."[47] Repetition of this procedure using ^{3}H-diazomethane supplies O-methyl-^{3}H-seco-loganin. *Catharanthus roseus* shoots provided with this tritiated seco-loganin yielded the radioactive indole alkaloids catharanthine (7), vindoline (6), perivine (27), ajmalicine (4a), and serpentine (4b), with the incorporations displayed in Table II. Zeisel demethylation of the ajmalicine and catharanthine proved that 94% and 96% respectively, of the total activities were

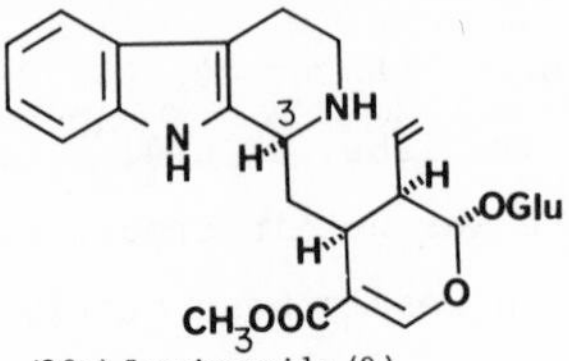

(28a) Isovincoside (?)

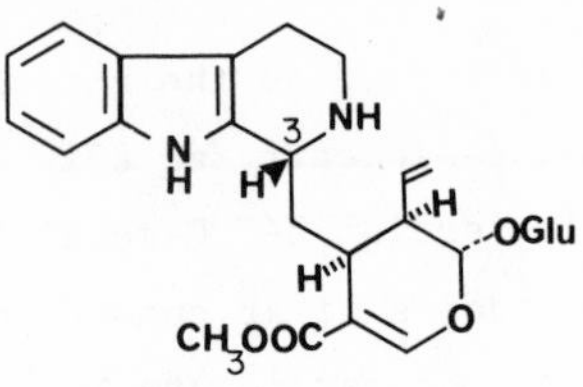

(28b) Vincoside (?)

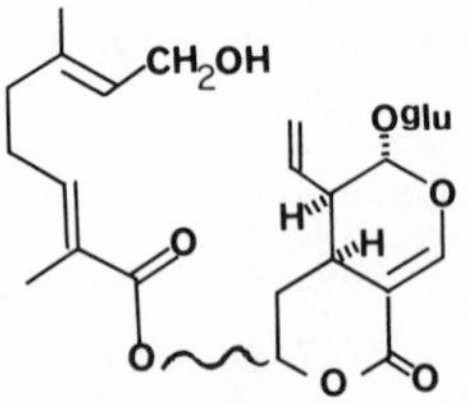

(29a) Foliamenthin

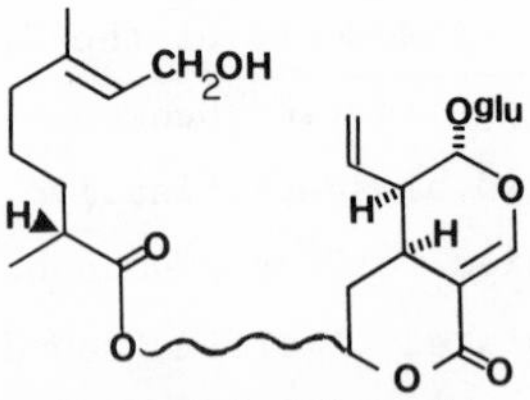

(29b) Dihydrofoliamenthin

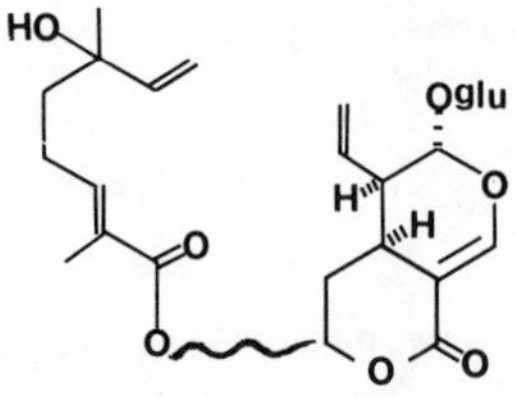

(30) Menthiafolin

located in the carbomethoxy functions. The vindoline yielded
desacetylvindoline by hydrolysis which contained 98% of the
original activity; following reduction of the desacetylvindoline,
a triol was produced carrying less than 1% of the former activity. The absence of significant transmethylation via the "C_1"
pool was apparent by the lack of activity of this triol, which
still contained the N-methyl and aryl-O-methyl groups of vindoline. The specific incorporation of seco-loganin into the three
skeletal classes of indole alkaloids is therefore demonstrated.

The seco-lactone, sweroside (31), also serves as a precursor
of indole alkaloids in $C.\ roseus.$[48] 10-^{14}C-Sweroside was shown

to be incorporated specifically into vindoline to the extent of
11%   an observation which may be explained by the initial transformation of (31) into seco-loganin:

(31)

As a prelude to the condensation of seco-loganin with tryptamine, the ability of *Catharanthus* plants to convert tryptamine
into the three classes of indole alkaloids was demonstrated.[49]
Tryptamine hydrochloride was then condensed with seco-loganin to
produce the hydrochlorides of the two epimeric β-carbolines (28a)
and (28b).[49]   The use of O-methyl-^{3}H-seco-loganin generated a
mixture of the correspondingly labeled β-carbolines which was
fed to *C. roseus*.   Incorporations into all three types of indole
alkaloids were observed (Table II).   Intact incorporations were
assured by the use of a mixture of doubly-labeled (28a) and (28b)
prepared from O-methyl-^{3}H-seco-loganin and U-^{3}H-tryptamine hydrochloride.   Degradations of the radioactive alkaloids produced
in this instance disclosed (Table II) that biosynthesis had
proceeded without significant alteration in the ratio of the
labels present in the two halves of the glucosides.[49]
    Partition chromatography has been utilized to separate (28a)
and (28b), which have been named "vincoside" and "isovincoside",
respectively.[49,50]   The stereochemistry assigned to C-3 of each
glucoside was the result of a correlation with ipecoside based
upon the molecular rotation differences method.   It is now known
from X-ray work[31] (*vide supra*) that the configuration at C-3 of

ipecoside is the opposite to that originally assigned, and it is
therefore highly probable that (28b) represents the correct structure of vincoside while (28a) corresponds to isovincoside. This
proves to be a most unexpected result as separation of the 0-
methyl-^{3}H-epimers, (28a) and (28b), and administration of them
individually to *C. roseus* leads only to the incorporation of
vincoside into the indole alkaloids.[49,50]   If, as it now appears,
vincoside bears a hydrogen of β-configuration at C-3, then an
inversion must occur at that carbon during the later stages of
biosynthesis which lead to those indole alkaloids having the
C-3 hydrogen atom in the α-configuration. The occurrence of
such an inversion is now clearly established for the biosynthesis of cephaeline (8) and emetine (9) in *Cephaelis ipecacuanha*.
These two alkaloids each possess a C-3 hydrogen of the α-configuration, but feeding experiments show that only desacetyl-
ipecoside, which must be assigned structure (32a) as a consequence of the X-ray analysis of ipecoside, serves as a precursor

(32a)Desacetylipecoside   (32b) Desacetylisoipecoside

of (8) and (9); the C-3 epimer, desacetyl-isoipecoside (32b) is
not incorporated.[51]   Furthermore, experimental evidence is available indicating that the inversion at C-3 is accomplished without loss of the hydrogen atom from this position.[51]
    The same situation probably obtains in *Catharanthus* plants,
if vincoside is correctly represented by (28b). A mixture of
5-[3]H- and 0-methyl-^{3}H-loganin has been administered to *C. roseus*

shoots, and specific incorporations into ajmalicine (4a) and serpentine (4b) demonstrated by removal of the methoxy group from the labeled alkaloids.[50] The radioactive ajmalicine carried the expected quantity of skeletal tritium, while the radioactive serpentine carried no skeletal tritium. These observations indicate: a) that the label from C-5 of loganin resides exclusively at C-3 of (4a), as anticipated, and b) that no loss of label from this position occurs during the biosynthesis of ajmalicine.

Evidence for the natural occurrence of seco-loganin, vincoside, and isovincoside stems from various sources. *C. roseus* plants to which 5-^{3}H-loganin had been administered were worked up with the addition of inactive samples of the aforementioned substances as carriers. The isolated seco-loganin fraction was treated with dopamine and thereby converted into ipecoside.[49] The activity of the ipecoside corresponded to greater than 6% incorporation of loganin into seco-loganin. Acetylation of the β-carboline fraction yielded the penta-acetates of vincoside and isovincoside, each of which possessed activities as their N-acetyl derivatives corresponding to 1.5% incorporation.[49] More direct evidence for the natural occurrence of vincoside and isovincoside has come from the isolation of vincoside penta-acetate from an acetylated extract of *C. roseus* seedlings,[52] and from the isolation of N-acetyl-vincoside from mature *C. roseus* plants.[49] In addition, an extract of *Rhazya stricta* Decaisne (Apocynaceae) has been reported to yield an amorphous glucoside ("strictosidine") which appears to be identical with isovincoside.[53]

The major outlines of the processess leading from mevalonic acid to vincoside having been revealed, the light of experiment is now being focused on the two remaining areas of biosynthetic darkness. These encompass the stages between geraniol and loganin, and the processes whereby vincoside is transformed into the three classes of indole alkaloids.

THE STAGES BETWEEN GERANIOL AND LOGANIN

The transformation of geraniol (14) and nerol (14, isomeric at 2, 3 double-bond) into the C_{10} skeleton of loganin requires oxidation of the C-9 and C-10 methyl groups, oxidation of C-1 to the aldehyde level, saturation of the Δ^2-double-bond, and formation of the cyclopentane ring. The isolation of the monoterpenes (29a, b) and (30), which are oxidized at C-10, from *M. trifoliata*[45,46] suggested that oxidation of C-10 might be the initial step. A series of labeled geraniol derivatives of this type was therefore prepared and each compound (32c-40) fed to *C. roseus*.[54,55] Only (32c) and (33) produced significant incorporation figures (Table IIIa). Administration of sodium 2-^{14}C-mevalonate to *C. roseus* had previously shown[20,21,22] (*vide supra*) that equilibration of C-2 and C-6 of one mevalonate-derived unit occurs by the time these carbons have arrived in the final al-

TABLE IIIa

Incorporation of Geraniol Derivatives into Loganin and *Catharanthus* Alkaloids

Precursor	Loganin (26)	Ajmalicine (4a)	% Incorporations Serpentine (4b)	Catharanthine (7)	Vindoline (6)	Perivine (27)
$1-{}^3H_2$-(32) and (33)	0.31	0.36	0.20	0.36	0.25	0.021
$9-{}^{14}C$-(32)	0.09	0.17		0.5	0.72	
$9-{}^{14}C$-(33)	0.16	0.14		0.9	1.2	
$9-{}^{14}C$-(34)		0.0004		0.002	0.003	
$1-{}^3H_2$-(35)	< 0.024	< 0.001	< 0.0025	0.020	0.017	0.0006
$9-{}^{14}C$-(35)	0.0002			0.002	0.001	
$1-{}^3H_2$-(36)		< 0.005	< 0.0034	0.011	0.011	0.0007
$10-{}^3H_2$-(37)		< 0.002	< 0.0048	0.032	0.048	0.0014
$1-{}^3H_1$-(38)		0.0				
$1-{}^3H_2$-(39)	0.0002	< 0.002 0.0004		0.0004	0.0008	
$1-{}^3H_2$-(40)	< 0.003	0.0	0.0007	0.0008	0.0006	0.0001

kaloids. Retention of their individuality as far as the iso-
propylidene moiety of geraniol appears likely on the basis of
a) the results of a study on the stereochemistry of label incorporation from $4-{}^{3}$H, $2-{}^{14}$C-(3R, 4R) mevalonate into *Catharanthus* alkaloids (*vide infra*)[56] and b) the observation that
this individuality is preserved in the corresponding segment of
the triterpene ursolic acid which is produced simultaneously in
the same plant.[57] It is now apparent that this equilibration
takes place during the conversion of (32c) and (33) into loganin,
since degradation of the ^{14}C-loganin and ^{14}C-labeled alkaloids
from the feeding of (32c) and (33) found *ca.* 50% of the label
in the methoxycarbonyl groups (Table IIIb).[55]

The data above, combined with the observation that iridodial
(41) is not significantly incorporated into vindoline by *C.
roseus*,[58] leads one to suspect that oxidation of both C-9 and
C-10 of geraniol occurs before formation of the cyclopentane ring.
If a trialdehyde such as (42) were involved, then cyclization
could generate a cyclic trialdehyde (43), and thereby equili-
brate C-9 and C-10.

(41) Iridodial   (42)   (43)

No information is presently available on these stages and
the only other intermediate in the geraniol to loganin pathway
whose identity seems secure is deoxyloganin (44). Structural
relationships provided the clue that the insertion of the C-5
hydroxyl group might be the penultimate step in loganin biosynthesis. In order to test this hypothesis, O-methyl-^{3}H-deoxy-

TABLE IIIb

Labeling Patterns from Feeding of $9-{}^{14}C$-10-Hydroxy-Geraniol and -Nerol

Precursor	Loganin (26)	% of Total Activity in Methoxycarbonyl Group		
		Ajmalicine (4a)	Catharanthine (7)	Vindoline (6)
$9-{}^{14}C-(32)$	39.8	39.4	43.0	38.8
$9-{}^{14}C-(33)$	38.5	48	43.3	39.8

loganin was partially synthesized from loganin by the route
shown (Scheme VI).[59] It was then administered to *C. roseus*
shoots and led to the incorporations of Table IV. Specific
labeling was assured by Zeisel demethylation of loganin and of
three of the alkaloids (Table IV), and by conversion of vindoline first to desacetylvindoline (97% of original activity) and
then to a triol by reduction (1% of the original activity). As
further evidence, deoxyloganin was detected in *M. trifoliata* and
C. roseus plants by dilution analysis after the shoots had ab-
sorbed 1-^{3}H-geraniol. A mixture of the two O-methyl-^{3}H-labeled
olefins (45) and (46) was also fed to *C. roseus,* leading to poor
incorporations (Table IV). This result indicates that the hydroxylation of deoxyloganin is accomplished by the direct insertion of the hydroxyl function rather than by addition of
water to a double-bond. Very poor incorporations were also the
result of feeding O-methyl-^{3}H-deoxyloganin aglucone (from treatment of labeled deoxyloganin with emulsin) to *C. roseus,* behavior
which may be attributed either to a penetration problem or to
the possibility that the link between the sugar and the terpene
skeleton is forged at an earlier stage.[59] The importance of
deoxyloganin as a general precursor of the iridoid glucosides
is emphasized by the reported ability of deoxyloganic acid (44,
ester hydrolyzed) to serve as the precursor of verbenalin (23),
aucubin (47), and asperuloside (48) as well as of loganin.[60]

     The stereochemistry of the hydroxylation of deoxyloganic
acid to loganic acid (26, ester hydrolyzed) in *Swertia caroli-
nenses* (Walt.) Ktze. (Gentianaceae) has been subjected to scru-

(47) Aucubin   (48) Asperuloside

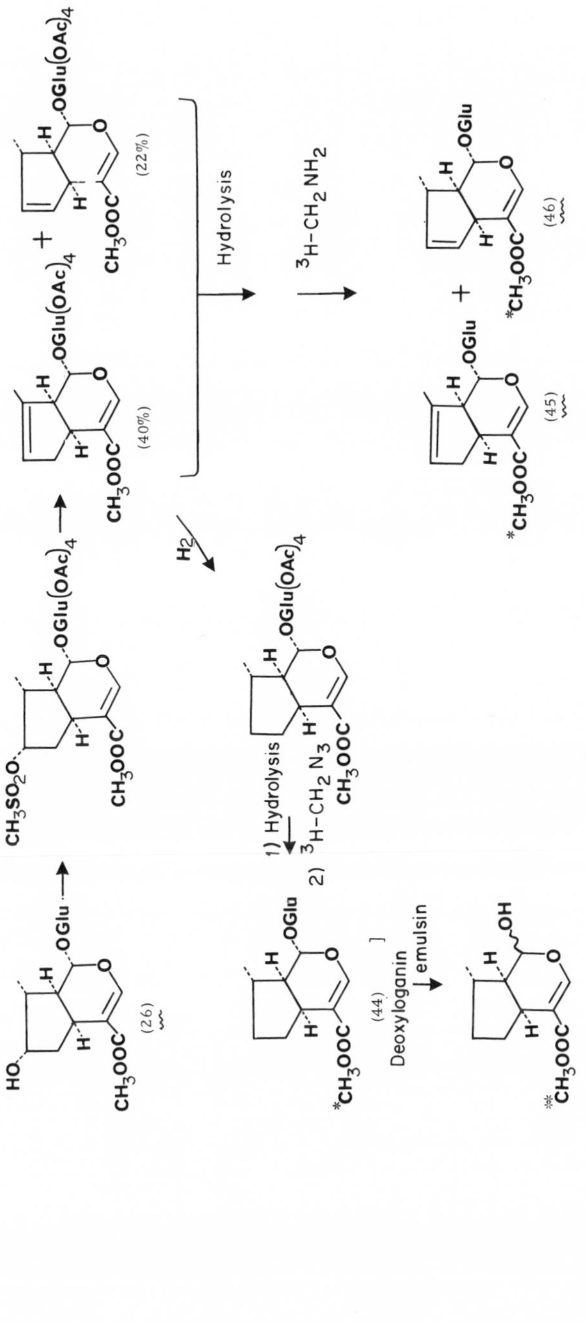

Scheme VI

TABLE IV

Administration of Loganin Derivatives to *Catharanthus roseus*

Precursor	Loganin (26)	Ajmalicine (4a)	% Incorporations Serpentine (4b)	Catharanthine (7)	Vindoline (6)	Perivine (27)
O-Me-^{3}H-Deoxy-loganin (44)	6.4	0.10	0.51	0.29	0.24	0.015
% of Total Activity at Methoxyl	99	98	97	97		
O-Me-^{3}H-Olefins (45) and (46)	< 0.001	< 0.001	< 0.001	< 0.001	< 0.001	< 0.001
O-Me-^{3}H-Deoxy-loganin Aglucone	< 0.08	< 0.001	< 0.001	< 0.001	< 0.001	< 0.001

tiny.[61] Administration of 2-^{3}H, 2-^{14}C (3R, 4R) and 2-^{14}C (3R, 4S)-mevalonate to this plant leads to the conclusion that the pro-R hydrogen at C-2 of mevalonic acid is retained at C-5 of loganic acid whereas the pro-S hydrogen is lost as a result of the hydroxylation process. If the synthesis of geraniol from mevalonate in *Swertia* follows the established stereochemical course,[62] then the loss of the pro-S hydrogen from C-5 corresponds to hydroxylation at that center with retention of configuration. This is the stereochemical consequence which most often results from hydroxylation at a saturated carbon atom.[63]

THE STAGES BEYOND VINCOSIDE

The biological conversion of vincoside (28a or b) into the three major classes of indole alkaloids may be imagined to proceed by removal of the sugar to give vincoside aglucone (28a or b, sugar hydrolyzed off), which should be in equilibrium with the ring-opened aldehyde (49), followed by condensation of the unmasked aldehyde function with $N_{(b)}$ to generate an iminium salt (50); if vincoside is in fact represented by (28b), then epimerization at C-3 could occur at this stage *via* reversible ring-opening of (50) to (51); reduction of the epimerized iminium salt could then yield either corynantheine aldehyde (52) or geissoschizine (53) (Scheme VII). Experimental evidence is now available bearing on the role that each of the two chemical protagonists, (52) and (53), plays in the biosynthesis of the *Catharanthus* alkaloids.

Curiously, the importance of corynantheine aldehyde in the biochemical economy of *Catharanthus* appears to vary with the age of the plants being examined. When O-methyl-^{3}H-corynantheine aldehyde was infused into *C. roseus* seedlings, radioactive catharanthine (0.3% incorpn.), and vindoline (0.1% incorpn.) were produced.[64] In contrast, when mature *Catharanthus* plants were supplied with O-methyl-^{3}H- and ring C-^{3}H-corynantheine al-

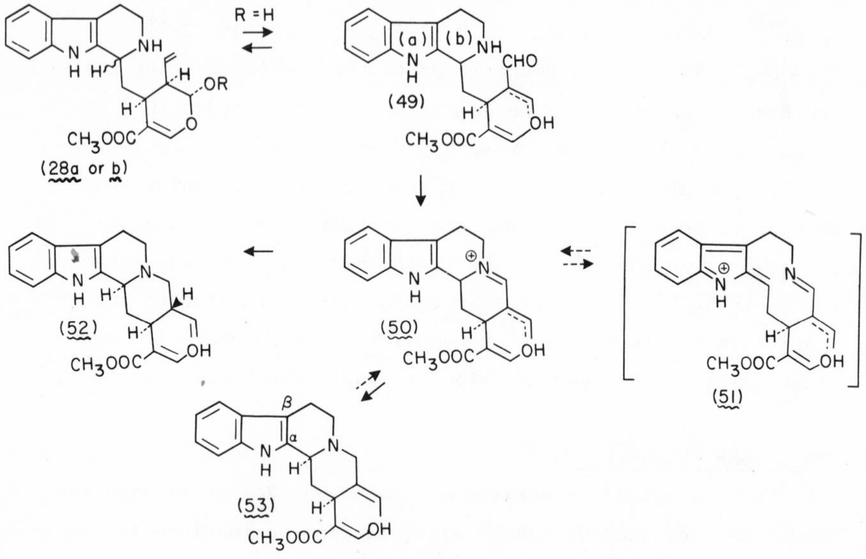

Scheme VII

dehyde, much lower incorporations into the indole alkaloids were the result (see Table V).[56,64]

On the other hand, feeding experiments with labeled geisso-schizine demonstrate that it will serve as an alkaloid precursor in both seedlings and in mature plants. Ar-^{2}H-geissoschizine, when fed to the seedlings,[65] leads, *inter alia*, to deuterated akuammicine (5) and coronaridine (54), as analyzed by mass spec-tral methods. Both Ar-^{3}H- and O-methyl-^{3}H-geissoschizine have been prepared from geissospermine (55), and supplied to mature *C. roseus* plants.[66] Radioactive ajmalicine, serpentine, akuam-micine, vindoline, and catharanthine were produced (see Table V). A double - labeling experiment (Table V) verified intact in-corporations and blank experiments ruled out the *in vitro* con-version of geissoschizine into (4a,b) and (5) under the condi-tions of workup. The presence of geissoschizine in *Catharanthus* plants was then demonstrated by both isotopic dilution after

TABLE V

Incorporations of Vincoside, Corynantheine Aldehyde, and Geissoschizine into *Catharanthus* Alkaloids

Precursor	Ajmalicine (4a)	Serpentine (4b)	% Incorporations Akuammicine (5)	Catharanthine (7)	Vindoline (56)
O-Me ^{3}H-Corynantheine Aldehyde (52)	< 0.001			0.3* < 0.001	0.1* 0.003
Ar-^{2}H-Geissoschizine (53) (a) (b)			1.53* 5.0*	0.35*‡	
O-Me-^{3}H-Geissoschizine	0.22			0.41	0.35
Ar-^{3}H-Geissoschizine (c) (d)	0.12 0.11	0.58 0.65	0.63 0.84	0.21 0.31	0.13 0.14
Ar-^{3}H, O-Me-^{3}H-Geissoschizine	0.12	0.82	2.0	0.47	0.41
Ar-^{3}H/O-Me-^{3}H = 2.25	2.1	2.1	2.25	2.0	1.7
Ar-^{3}H-Vincoside and C-3 Epimer (28a or b)	0.95	1.6	0.76	0.35	0.11

(a) = 10 mg precursor/100 gm seeds
(b) = 200 mg precursor/100 gm seeds
(c) = } two methods of precursor purification
(d) = }

* Seeds

‡ As Coronaridine (54)

(54) Coronaridine

(55) Geissospermine

feeding 5-^{3}H-loganin (1.3% incorpn.) and by isolation of (53) from a large scale extraction.[66] The discovery of the efficient incorporation of geissoschizine into akuammicine is of particular interest since it demonstrates an α to β rearrangement to the *Strychnos* system. This rearrangement generates the bond between C-2 and C-16 which is present in the *Aspidosperma* and Iboga alkaloids. As might be expected, vincoside also serves as a precursor of akuammicine in addition to being incorporated into ajmalicine and serpentine (Table V). The higher incorporation of vincoside into (4a, b) relative to that of geissoschizine is readily rationalized on the basis that vincoside may proceed directly to (4a, b) *via* the iminium salt (50), while (53) must presumably be reoxidized to (50) before being transformed into ajmalicine and serpentine.

A clue to the nature of the processes leading to the rearrangement of the *Corynanthe-Strychnos* unit (1) to the *Aspidosperma* (2) or Iboga (3) units was provided by attempts to simulate the rearrangements *in vitro*. It was reported[67] that treatment of the *Aspidosperma* base, (-)-tabersonine (56), with refluxing acetic acid led to the Iboga alkaloids (±)-catharanthine (7) and (±)-pseudocatharanthine (57), and similarly, that the same treatment of (+)-stemmadenine (58, *Corynanthe* skeleton) produced (±)-tabersonine, (±)-catharanthine, and pseudocatharanthine.

These interconversions were rationalized as proceeding through the optically inactive acrylic ester (59), which was postulated to be the branching point in the pathway converting the *Cory-nanthe* skeleton into either the *Aspidosperma* or Iboga skeletons. The conversion of stemmadenine into (59) was rationalized[67] as proceeding by isomerization of the exocyclic double-bond into the 20, 21-position followed by fragmentation:

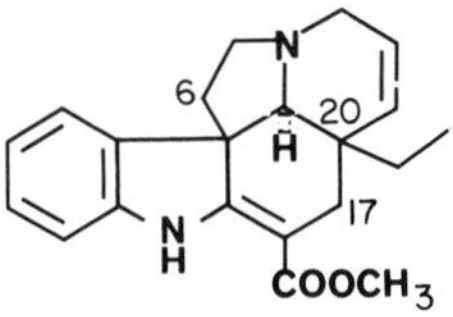

(56) Tabersonine

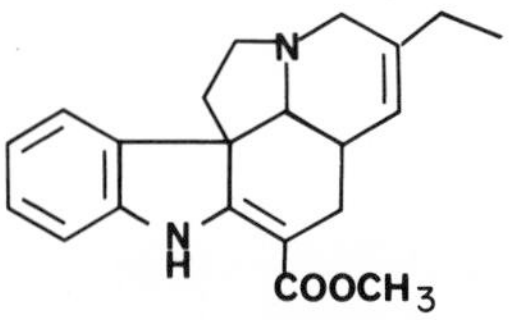

(57) Pseudocatharanthine

Evidence suggesting that such a mechanism might obtain *in vivo* comes from the reported failure[50] of O-methyl-^{3}H-dihydrovinco-side (28a or b, with vinyl group saturated) to serve as an al-kaloid precursor in *C. roseus*. Other evidence that is avail-able (*vide infra*) also strengthens the supposition that (59) plays a crucial role in the *in vivo* rearrangement processes, but the *in vitro* experiments which led to the adumbration of (59) appear to have been effectively refuted. A careful reexam-ination of the result of treating (56) and (58) with hot acetic acid failed[68] to confirm the original report. Stemmadenine was heated in anhydrous acetic acid under a variety of conditions

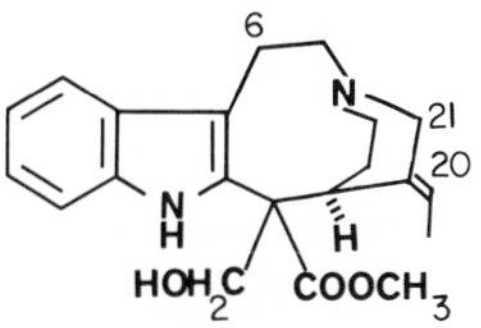

(58) Stemmadenine

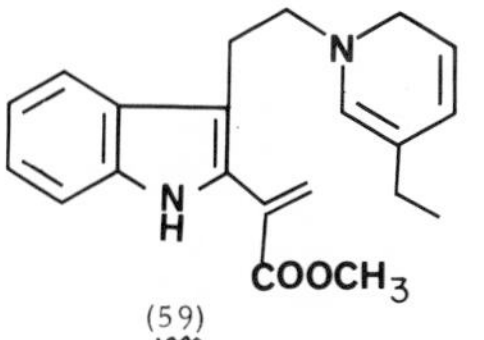

(59)

and found to give back only the starting alkaloid (*ca*. 15 %)
and its O-acetyl derivative (*ca*. 50 %). The reaction mixture
was checked using four different tlc systems and no (7), (56),
or (57) was detected. When the reaction was repeated with O-^{3}H-
acetic acid which should lead to labeled (7) and (57), and the
mixture of products analyzed by dilution analysis, a maximum
possible yield of 4 x 10^{-3}% of (57) was indicated. The acetic
acid treatment of tabersonine was found to yield unchanged alkaloid (48 %), dihydrotabersonine (14 %), and three closely-re-
lated new compounds, the principal one of which was an isomer of
catharanthine, called "allocatharanthine". This new base (17 %
yield) gave a mass spectrum identical to that of catharanthine
and exhibited similar tlc behavior to catharanthine. Spectral
data allowed the assignment of structure (60) to allocatharanthine. The two other new products were dihydroallocatharanthine
(15 %) and acetoallocatharanthine (6 %) (61). The formation of
(60) from tabersonine does not require the intermediacy of the
acrylic ester (59), and probably proceeds by the following mechanism, without fission of the 17, 20-bond:

(60) Allocatharanthine

Distinctly different reaction paths intervene when tabersonine, catharanthine, and pseudocatharanthine are heated neat
or in an inert solvent (xylene or methanol).[68,69] These *in
vitro* conversions, which are also postulated to proceed through
the acrylic ester (59), are summarized in Scheme VIII. The isolation of the pyridinium salt (62) is of particular interest.

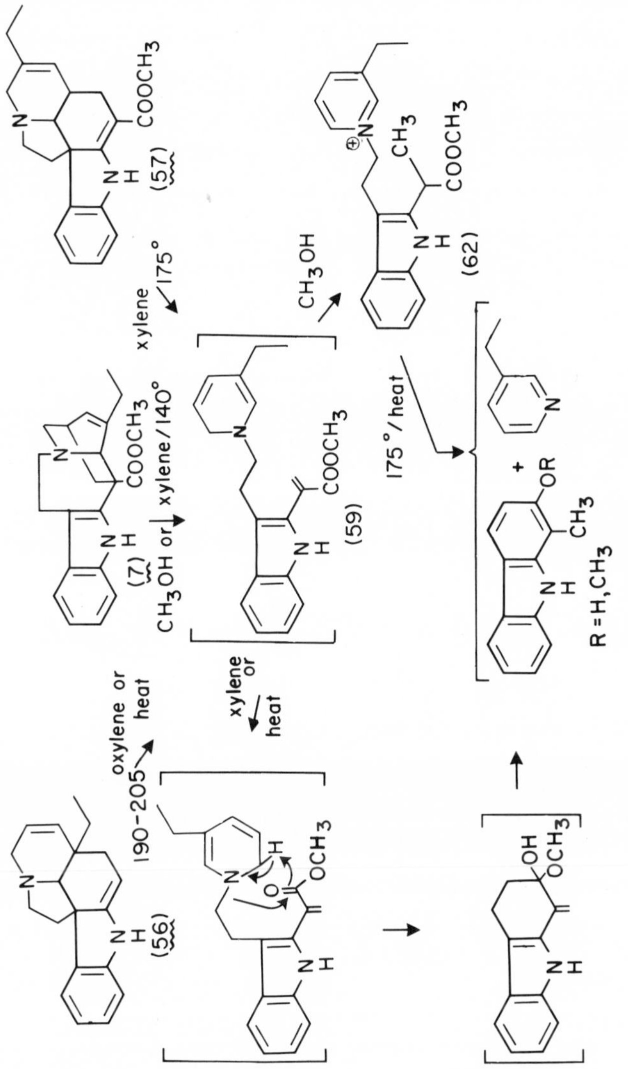
(57)
(7)
(56)
(59)
(62)
xylene 175°
CH_3OH or xylene/140°
CH_3OH
oxylene or heat 190-205°
xylene or heat
175°/heat
R = H, CH_3

Scheme VIII

Indirect evidence for the presence of (59) as a fugative intermediate in plant tissues derives from various sources. The acrylic ester derivative (63, R = OH) has been synthesized and detected in *Rhazya orientalis* shoots after the feeding of 0-methyl1-^{3}H-loganin (0.013% incorpn.).[70] When the same experiment was conducted with *C. roseus*, radioactive (63, R = OH) was again isolated, but it was of "very low specific activity".

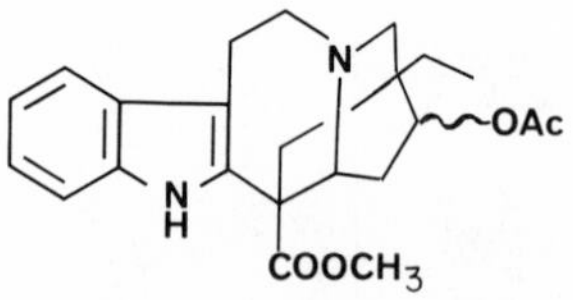

(61) Acetoallocatharanthine 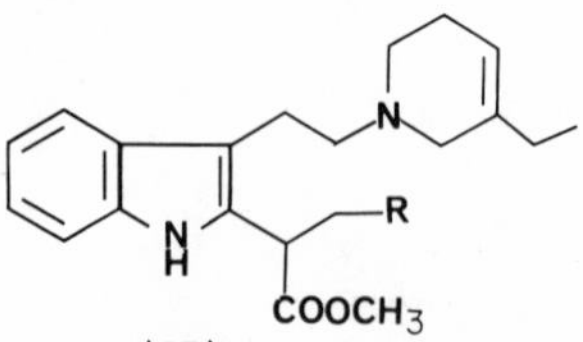   (63)

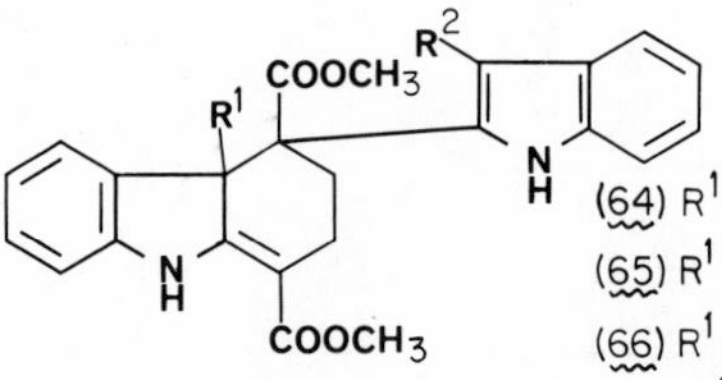

(64) R^1 = R^2 = X

(65) R^1 = R^2 =$_2$Y

(66) R^1 = X, R^2 =Y

or

R^1 = Y, R^2 = X

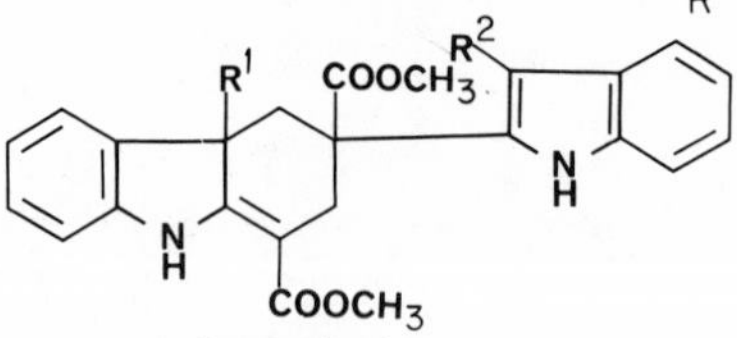

Presecamines

X = −CH$_2$CH$_2$N

Y = −CH$_2$CH$_2$N

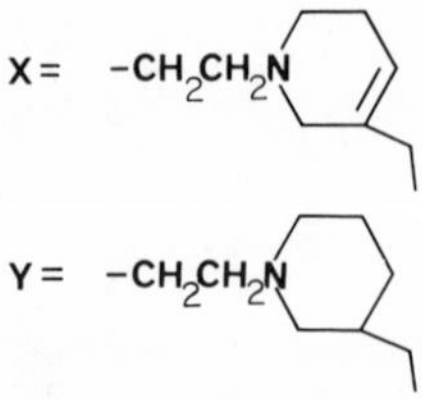

The completely reduced acrylic ester derivative (63, R = H,
double-bond in pyridine ring reduced) has also been synthesized
and it was detected in *Rhazya orientalis* after the feeding of
2-^{14}C-tryptophan (0.5% incorpn.).[71] Finally, *Rhazya* plants
have been found to contain[72] a family of indole alkaloids
("presecamines"), structures (64 - 66), which appear to be dimers
of the putative acrylic ester (59).

Considerable information on the stages of *Catharanthus* alkaloid biosynthesis which lie beyond vincoside has been gathered
by examination of the sequential formation of alkaloids in ger-
minating *C. roseus* seeds.[52,65] The dry seeds are themselves
devoid of alkaloids, but after 10-12 days of germination, the
mixture of indole alkaloids formed closely resembles that present in mature plants. The onset of alkaloid production is
detectable by tlc after *ca.* 24-26 hours, and vincoside, ajmalicine, and corynantheine (21) are present at this time. Over
the 28 to 40 hour period, corynantheine aldehyde (52) and geissoschizine (53) appear, to be followed by stemmadenine (58)
after *ca.* 50 hours. Each of these alkaloids carries the *Corynanthe* unit. The *Aspidosperma*-type alkaloid tabersonine (56)
also puts in an appearance after 50 hours, and the Iboga-type
alkaloids catharanthine (7) and coronaridine (54) arrive on the
biosynthetic scene after 100-160 hours of germination time have
elapsed. The *Aspidosperma*-type base vindoline (6) finally
reaches detectable concentration after about 200 hours. This
isolation sequence is summarized in Table VI. This Table contains four additional entries which correspond to the appearance of four hitherto unknown alkaloids. Three of these bases,
the β-hydroxyindolenine (67), the diol (68), and geissoschizine
oxindole (69) are reported to be detectable after about 28 to
40 hours of germination. They were assigned the structures indicated,[52] but no experimental evidence has yet been provided. If
the proposed structures are correct, they delineate the stages

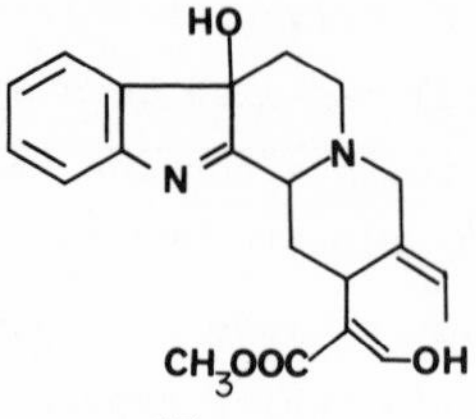

(67)

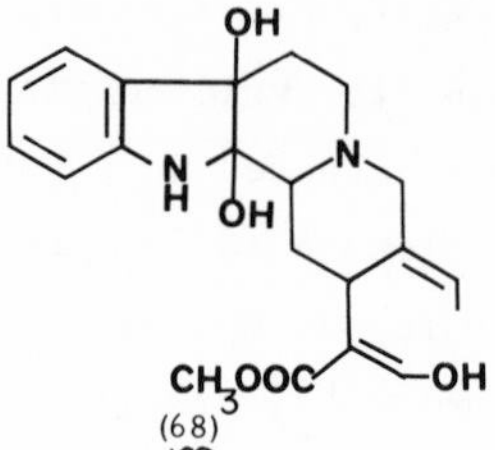

(68)

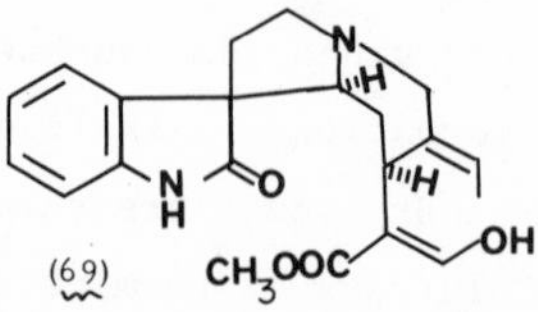

(69)   CH₃OOC

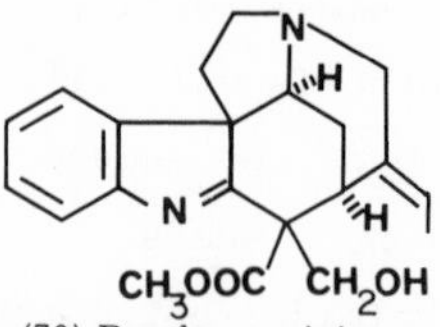

(70) Preakuammicine

by which geissoschizine undergoes the α to β rearrangement required to arrive at the *Strychnos* alkaloids. The forging of
the remaining link in the chain of biochemical events leading
from vincoside to stemmadenine was accomplished[73] by the isolation of the fourth new alkaloid listed in Table VI. The experimental data favor structure (70) for this substance, which
therefore corresponds to the long sought "C_{10}-*Strychnos*" alkaloid. (70) has been dubbed "preakuammicine", since storage in
solution at room temperature transforms it into akuammicine.
Treatment with sodium borohydride converts (70) into stemmadenine (Scheme IX).

The isolation sequence displayed in Table VI clearly favors
the order of alkaloid biosynthesis, *Corynanthe* to *Corynanthe-
Strychnos* to *Aspidosperma* to Iboga. Additional evidence favor-
ing this order of biosynthetic events derives from a series of
alkaloid feedings,[52,64] the results of which are summarized in
Table VII. It has been reported that geissoschizine oxindole
(69) can be partially synthesized from geissoschizine,[52] but
experimental detail is lacking. When administered to *C. roseus*

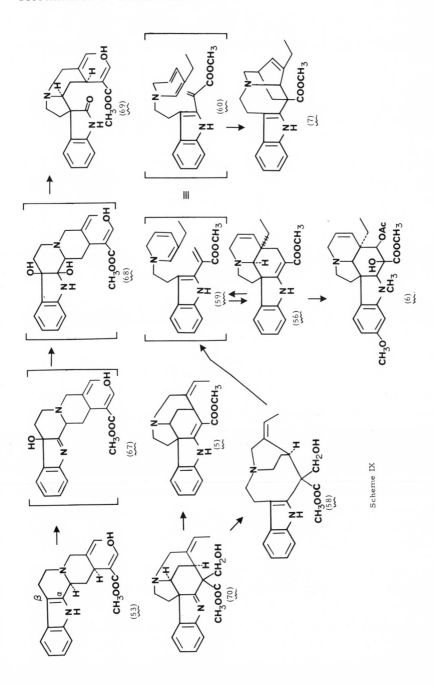

Scheme IX

TABLE VI

Isolation of Alkaloids from *Catharanthus roseus* Seedlings

Germination Time (Hours)	Alkaloid Isolated	Type
0	None	–
26	Vincoside (28a or b)	*Corynanthe*
	Ajmalicine (4a)	"
	Corynantheine (21)	"
28–40	Corynantheine Aldehyde (52)	"
	Geissoschizine (53)	"
	β-Hydroxyindolenine (67)	"
	Diol (68)	"
	Geissoschizine Oxindole (70)	"
40–50	Preakuammicine (70)	*Corynanthe-Strychnos*
	Akuammicine (5)	*Strychnos*
	Stemmadenine (58)	*Corynanthe-Strychnos*
	Tabersonine (56)	*Aspidosperma*
100–160	Catharanthine (7)	Iboga
	Coronaridine (54)	"
200	Vindoline (6)	*Aspidosperma*

TABLE VII

Administration of Labeled Indole Alkaloids to *Catharanthus roseus* Seedlings

Precursor	Ajmalicine (4a)	Serpentine (4b)	Akuammicine (5)	% Incorporations Catharanthine (7)	Vindoline (6)	Tabersonine (56)
1. Ar-^{3}H-Geisso-schizine oxindole (69)			0.55		0.05	
2. Ajmalicine; Ring C-^{3}H		> 1.8[a]		0.007[a] 0.3	0.004[a] 0.6	
^{3}H-OMe					0.001[a]	
3. O-^{3}H-Me, 6-^{14}C-Stemmadenine (58)				0.30	0.95	0.10
O-^{3}H-Me/^{14}C= 92.8/7.2				91.8/8.2	91.9/8.1	92.3/7.7
4. O-^{3}H-Me, 6-^{14}C-Tabersonine (56)				0.14	1.10	
O-^{3}H-Me/^{14}C= 95.8/4.2				95.6/4.4	96.0/4.0	
5. O-^{3}H-Me-Catharanthine					0.001	< 0.001

[a] mature plant

seedlings, this partially synthetic oxindole was apparently
incorporated efficiently into akuammicine and moderately well
into vindoline (entry 1, Table VII). Rather surprisingly, *Catharanthus* seedlings are able to incorporate ajmalicine into alkaloids of both the *Aspidosperma* and Iboga groups,[64] but this cannot be accomplished by mature plants (entry 2, Table VII).[56]
This behavior may be attributed to the presence in the seedlings
of enzymes which allow the conversion of ajmalicine back to
geissoschizine:

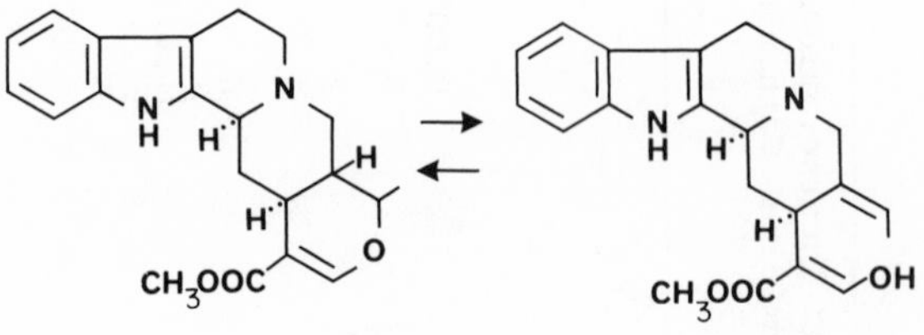

Entries 3 - 5 of Table VII demonstrate the expected ability
of stemmadenine and tabersonine to serve as precursors of vindoline and catharanthine, and more significantly, show that
catharanthine will not serve as an effective precursor of either
tabersonine or vindoline. The biosynthetic sequence stemmadenine to tabersonine to catharanthine is thereby suggested and
this harmonizes well with the sequential isolation data of
Table VI. It should be noted that the very high incorporation
of tabersonine into vindoline points to the introduction of the
11-methoxy, $N_{(a)}$-methyl, and acetoxy functions at a late stage
in the biosynthesis. The conversion of labeled tabersonine
into catharanthine has also been reported to occur in six month
old *Catharanthus roseus* plants.[74] On the basis of the data
contained in Tables VI and VII, a complete pathway may be for-
mulated leading from geissoschizine to the three major classes
of indole alkaloids that are found in *Catharanthus roseus*. This

pathway is delineated in Scheme IX. Brackets surround those
intermediates whose existence has yet to be convincingly proved.

It remains to discuss a group of experiments which were
designed to probe some of the mechanistic aspects of indole alkaloid biosynthesis by utilizing doubly-labeled precursors. The
results[56] of this series of *C. roseus* feedings are presented in
Table VIII, and they lead to the following conclusions. The administration of $1-{}^3H_2$, $2-{}^{14}C$-geraniol results in *ca.* 50 % retention of the tritium on conversion into loganin; oxidation of C-
1 of geraniol to the aldehyde level therefore appears to be a
stereospecific process. The resulting C-1 labeled loganin re-
tains its activity throughout the transformations leading to the
three alkaloid families; if stemmadenine (58) is converted into
the *Aspidosperma* and Iboga type alkaloids *via* the hypothetical
acrylic ester (59), then this result indicates that the loss of
the proton from C-21 of stemmadenine associated with the formation of (59) must be stereospecific. With $2-{}^3H$, $2-{}^{14}C$-geraniol,
no important loss of tritium occurs during the formation of loganin, but nearly complete loss takes place in the stages leading to the three alkaloid types. Thus, if saturation of the 2,
3-double-bond of geraniol is a step in its conversion to loganin, both the reduction and the subsequent loss of a proton
from C-2 on formation of the cyclopentane ring must be stereospecific. Further insight is provided by the incorporation of
$4-{}^3H$, $2-{}^{14}C$-(3R, 4R)-mevalonate. It is known that the (4R)-form
of this acid is converted *in vivo* to all *trans* isoprenoids with
unchanged tritium to carbon-14 ratios;[75] in accordance with this
information, complete retention of tritium is observed in the
biosynthesis of loganin from this form of labeled mevalonate,
while complete loss of tritium obtains with the (4S)-compound.
Approximately 50 % retention of the tritium of loganin labeled
by (4R)-mevalonate is associated with the processes leading to
each of the major classes of indole alkaloids. Since the tritium

TABLE VIII

Administration of Doubly-Labeled Geraniol, Nerol, and Mevalonate to *Catharanthus roseus*

Precursor	Loganin (26)	Ajmalicine (4a)	Serpentine (4b)	Catharanthine (7)	Vindoline (6)	Perivine (27)
			% Retention of ^{3}H Relative to ^{14}C			
$1-^3H_2$, $2-^{14}C-$Geraniol	45	44	43	48	47	49
$2-^3H$, $2-^{14}C-$Geraniol	95	<5	<5	<5	<5	<5
$2-^3H$, $2-^{14}C-$Nerol	101	<5	<5	<5	<5	<5
Sodium $4-^3H$, $2-^{14}C-$(3R,4R)-(±)-Mevalonate	109 98	46 47	– 49	56 42	57 47	– 50
Sodium $4-^3H$, $2-^{14}C-$(3R,4S)-(±)-Mevalonate	10±5	<5	<5	<5	<5	<5

at C-2 of loganin has already been proven to be removed during these steps, it follows that the tritium at C-7 of loganin must have been retained. Therefore, the configuration of loganin at C-7 determines the stereochemistry of the corresponding center of ajmalicine (C-15), and presumably of the other *Corynanthe-Strychnos* alkaloids as well. On the other hand, the stereochemical correspondence between C-2 of loganin and C-20 of ajmalicine is seen to be only coincidental, since the proton is lost from that position of loganin during the biosynthesis. These findings are also clearly incompatible with the proposal that swertiamarin (71) lies on the main route to the indole alkaloids.[48]

(71) Swertiamarin

REFERENCES

1. M. Hesse, Indolalkaloide in Tabellen, Springer-Verlag, Berlin; vol. I, 1964; vol. II, 1968.

2. The Alkaloids, Ed. by R.H.F. Manske, Academic Press, New York; vol. VIII, 1965; vol. XI, 1968.

3. A.R. Battersby, Quart. Rev. (London), 15, 259 (1961).

4. K. Mothes and H.R. Schütte, Anqew. Chem. Int. Ed., 2, 441 (1963).

5. E. Schlittler and W.I. Taylor, Experientia, 16, 244 (1960).

6. A.R. Battersby, Pure Appl. Chem., 14, 117 (1967).

7. E. Leete, Chem. Ind. (London), 1960, 692; J. Amer. Chem. Soc., 82, 6338 (1960).

8. R.B. Woodward, Nature (London), 162, 155 (1948).

9. E. Leete, S. Ghosal, and P.N. Edwards, J. Amer. Chem. Soc.,
    84, 1068 (1962).

10. a) E. Wenkert and N.V. Bringi, J. Amer. Chem. Soc., 81,
    1474 (1959); b) E. Wenkert, J. Amer. Chem. Soc., 84, 98
    (1962).

11. P.N. Edwards and E. Leete, Chem. Ind. (London), 1961, 1666.

12. E. Leete and S. Ghosal, Tetrahedron Lett., 1962, 1179.

13. E. Leete, A. Ahmad, and I. Kempis, J. Amer. Chem. Soc.,
    87, 4168 (1965).

14. A.R. Battersby, R. Binks, W. Lawrie, G.V. Parry, and B.R.
    Webster, Proc. Chem. Soc. (London), 1963, 369.

15. H. Goeggel and D. Arigoni, Experientia, 21, 369 (1965).

16. K. Stolle, D. Gröger, and K. Mothes, Chem. Ind. (London),
    1965, 2065.

17. J. D. Bu'Lock, The Biosynthesis of Natural Products, McGraw-
    Hill, New York, 1965.

18. A.R. Battersby, R. Binks, W. Lawrie, G.V. Parry, and B.R.
    Webster, J. Chem. Soc., 1965, 7459.

19. R. Thomas, Tetrahedron Lett., 1961, 544.

20. F. McCapra, T. Money, A.I. Scott, and I.G. Wright, Chem.
    Commun., 1965, 537.

21. H. Goeggel and D. Arigoni, Chem. Commun., 1965, 538.

22. A.R. Battersby, R.T. Brown, R.S. Kapil, A.O. Plunkett, and
    J.B. Taylor, Chem. Commun., 1966, 46.

23. D.A. Yeowell and H. Schmid, Experientia, 20, 250 (1964).

24. A.R. Battersby, R.T. Brown, J.A. Knight, J.A. Martin, and
    A.O. Plunkett, Chem. Commun., 1966, 346.

25. P. Loew, H. Goeggel, and D. Arigoni, Chem. Commun., 1966,
    347.

26. A.R. Battersby, R.T. Brown, R.S. Kapil, J.A. Knight, J.A.

27. E. S. Hall, F. McCapra, T. Money, K. Fukumoto, J. R. Hanson,
     B. S. Mootoo, G. T. Phillips, and A. I. Scott, Chem. Commun.,
     1966, 348.

28. E. Leete and S. Ueda, Tetrahedron Lett., 1966, 4915.

29. A. R. Battersby, B. Gregory, H. Spencer, J. C. Turner, M. M.
     Janot, P. Potier, P. Francis, and J. Levisalles, Chem. Commun.,
     1967, 219.

30. A. R. Battersby and B. J. T. Harper, J. Chem. Soc., 1959,
     1748.

31. O. Kennard, P. J. Roberts, N. Isaacs, F. H. Allen, W. D. S.
     Motherwell, K. H. Gibson, and A. R. Battersby, Chem. Commun.,
     1971, 899.

32. A. R. Battersby, R. Binks, and T. P. Edwards, J. Chem. Soc.,
     1960, 3474.

33. J. M. Bobbitt and K. P. Segebarth, "The Iridoid Glycosides
     and Similar Substances," in Cyclopentanoid Terpene Deriva-
     tives, Ed. by W. I. Taylor and A. R. Battersby, Marcel Dekker,
     Inc., New York, 1969.

34. G. Buchi and R. E. Manning, Tetrahedron, 18, 1049 (1962).

35. C. Djerassi, T. Nakano, A. N. James, L. H. Zalkow, E. J.
     Eisenbraun, and J. N. Shoolery, J. Org. Chem., 26, 1192
     (1961).

36. H. Inouye, T. Arai, and Y. Miyoshi, Chem. Pharm. Bull. (Tokyo),
     12, 888 (1964).

37. A. J. Birch and J. Grimshaw, J. Chem. Soc., 1961, 1407; K.
     Sheth, E. Ramstead, and J. Wolinsky, Tetrahedron Lett., 1961,
     394.

38. A. R. Battersby, R. T. Brown, R. S. Kapil, J. A. Martin, and
     A. O. Plunkett, Chem. Commun., 1966, 890.

39. A. R. Battersby, R. S. Kapil, and R. Southgate, Chem. Commun.,
     1968, 131; A. R. Battersby, E. S. Hall, and R. Southgate, J.
     Chem. Soc., C, 1969, 721.

40. S. Brechbuhler-Bader, C. J. Coscia, P. Loew, Ch. von Szcze-
    panski, and D. Arigoni, Chem. Commun., 1968, 136.

41. H. Inouye, T. Yoshida, and S. Tobita, Tetrahedron Lett.,
    1968, 2945.

42. A. R. Battersby, R. S. Kapil, J. A. Martin, and L. Mo, Chem.
    Commun., 1968, 133.

43. P. Loew and D. Arigoni, Chem. Commun., 1968, 137.

44. A. R. Battersby and B. Gregory, Chem. Commun., 1968, 134.

45. P. Loew, Ch. von Szczepanski, C. J. Coscia, and D. Arigoni,
    Chem. Commun., 1968, 1276.

46. A. R. Battersby, A. R. Burnett, G. D. Knowles, and P. G.
    Parsons, Chem. Commun., 1968, 1277.

47. A. R. Battersby, A. R. Burnett, and P. G. Parsons, Chem. Com-
    mun., 1968, 1280; J. Chem. Soc., C, 1969, 1187.

48. H. Inouye, S. Ueda, and Y. Takeda, Tetrahedron Lett., 1968,
    3453.

49. A. R. Battersby, A. R. Burnett, and P. G. Parsons, Chem. Com-
    mun., 1968, 1282; J. Chem. Soc., C, 1969, 1193.

50. A. R. Battersby, A. R. Burnett, E. S. Hall, and P. G. Parsons,
    Chem. Commun., 1968, 1582.

51. A. R. Battersby and R. J. Parry, Chem. Commun., 1971, 901.

52. A. I. Scott, Accnts. Chem. Res., 3, 151 (1970).

53. G. N. Smith, Chem. Commun., 1968, 912; A. R. Battersby and
    G. N. Smith, unpublished results.

54. A. R. Battersby, S. H. Brown, and T. G. Payne, Chem. Commun.,
    1970, 827.

55. S. Escher, P. Loew, and D. Arigoni, Chem. Commun., 1970, 823.

56. A. R. Battersby, J. C. Byrne, R. S. Kapil, J. A. Martin,
    T. G. Payne, D. Arigoni, and P. Loew, Chem. Commun., 1968,
    951.

57. R. Giger, L. Botta, and D. Arigoni, unpublished work cited
    in ref. 55.

58. R. M. Bowman and E. Leete, Phytochemistry, 8, 1003 (1969).

59. A.R. Battersby, A.R. Burnett, and P.G. Parsons, Chem. Commun., 1970, 826.

60. H. Inouye, S. Ueda, Y. Aoki, and Y. Takeda, Tetrahedron Lett., 1969, 2351.

61. C.J. Coscia, L. Botta, and R. Guarnaccia, Arch. Biochem. Biophys., 136, 498 (1970).

62. J.W. Cornforth, R.H. Cornforth, G. Popjak, and L. Yengoyan, J. Biol. Chem., 241, 3970 (1966).

63. W. Charney and H.L. Herzog, Microbial Transformations of Steroids, Academic Press, New York, 1967.

64. A.A. Qureshi and A.I. Scott, Chem. Commun., 1968, 948

65. A.I. Scott, P.C. Cherry, and A.A. Qureshi, J. Amer. Chem. Soc., 91, 4932 (1969).

66. A.R. Battersby and E.S. Hall, Chem. Commun., 1969, 793.

67. A.A. Qureshi and A.I. Scott, Chem. Commun., 1968, 945.

68. R.T. Brown, J.S. Hill, G.F. Smith, K.S.J. Stapleford, J. Poisson, M. Muquet, and N. Kunesch, Chem. Commun., 1969, 1475.

69. A.I. Scott and P.C. Cherry, J. Amer. Chem. Soc., 91, 5872 (1969).

70. A.R. Battersby and A.K. Bhatnagar, Chem. Commun., 1970, 193.

71. R.T. Brown, G.F. Smith, K.S.J. Stapleford, and D.A. Taylor, Chem. Commun., 1970, 190.

72. G.A. Cordell, G.F. Smith, and G.N. Smith, Chem. Commun., 1970, 191.

73. A.I. Scott and A.A. Qureshi, J. Amer. Chem. Soc., 91, 5874 (1969).

74. J.P. Kutney, W.J. Cretney, J.R. Hadfield, E.S. Hall, V.R. Nelson, and D.C. Wigfield, J. Amer. Chem. Soc., 90, 3566 (1968).

75. G. Popjak and J.W. Cornforth, Biochem. J. (London), 101, 553 (1966).

76.   A.R. Battersby and K.H. Gibson, Chem. Commun. 902 (1971).

77.   A.I. Scott, J. Am. Chem. Soc. 94, 8262 (1972).

78.   A.I. Scott and C.C. Wei, J. Am. Chem. Soc., 94, 8264, 8265,
      8266 (1972).

79.   J.P. Kutney, J.F. Beck, N.J. Eggers, H.W. Hanssen, R.S. Sood,
      and N.D. Westcott, J. Am. Chem. Soc., 93, 7324 (1971).

80.   A.I. Scott, P.B. Riechardt, M.B. Sleytor and J.G. Sweeney,
      Bioorg. Chem., 1, 157 (1971)

ADDENDUM

The pace of research in the field of *Catharanthus* alkaloid
biosynthesis has decelerated since this review was written. The
addendum outlines recent developments, surveying the literature
to June 1973.

Additional evidence bearing on the retention of the C-3
hydrogen atom of vincoside (28b) during the formation of indole
alkaloids has been reported.[76] [5-^{3}H]-Loganin was administered
to *Catharanthus roseus* and radioactive ajmalicine (4a), vindoline
(6), and catharanthine (7) were isolated; incorporations were
0.2, 0.82, and 1.2%, respectively. Proof that the tritium label
was still located at the C-3 derived carbon atom of each alkaloid
was obtained by oxidative degradations. Dehydrogenation of
ajmalicine with mercuric acetate afforded dehydroajmalicine (72)
which was purified and reduced back to ajmalicine with borohydride; the recovered ajmalicine carried 0.2% of the original
activity in agreement with tritium labeling at C-3. In conjunction with related experiments examining the incorporation of
[5-^{3}H]-loganin into the ipecac alkaloids, it had been shown that
complete exchange of the hydrogens on the carbons adjacent to
C-3 by iminium-enamine equilibrium does not take place under the
conditions of mercuric acetate oxidation.[51] The loss of label
associated with the oxidation of ajmalicine to (72) cannot therefore be due to exchange of a label present at these positions
due to a biological 1,2-H shift. Oxidation of the labeled
catharanthine with iodine-sodium bicarbonate gave the neutral
lactam (73) which carried 5.3% of the original activity; since
exchange cannot take place from the bridgehead position (C-14)
under the reaction conditions, 95%, of the original tritium label
is located at C-3. Chromium trioxide-pyridine converted labeled
vindoline into a mixture of products, one of which had structure
(74) and retained 4.7% of the starting activity.

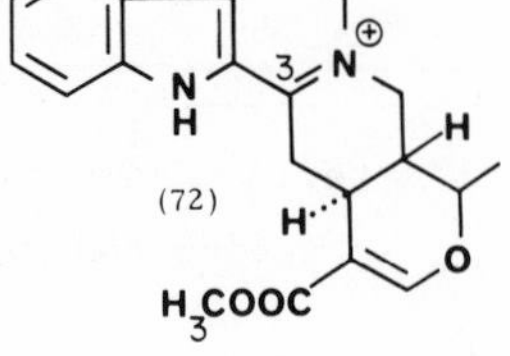

(72)

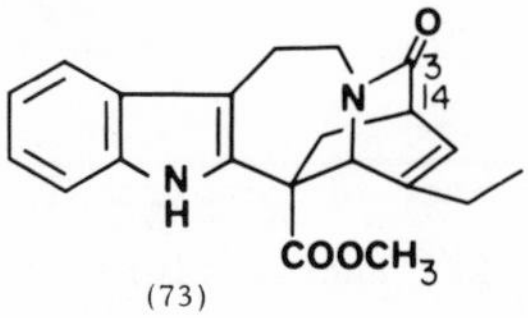

(73)

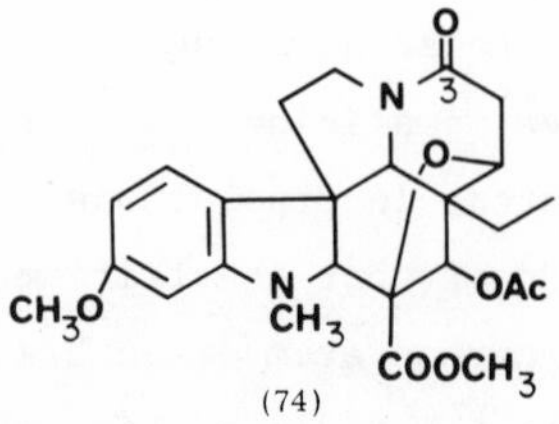

(74)

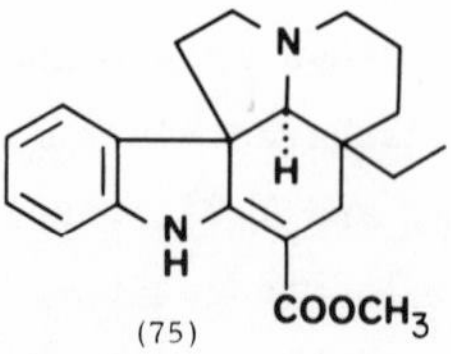

(75)

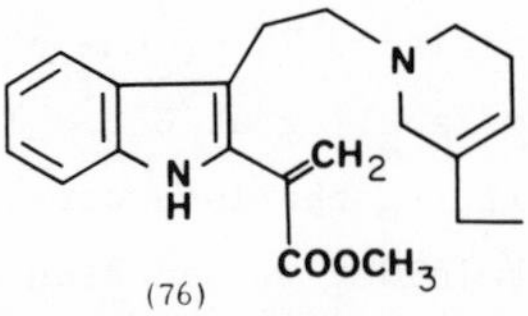

(76)

The controversy over the behavior of tabersonine (56) and stemmadenine (58) in refluxing acetic acid continues. In a series of recent communications[77-78], Scott reaffirms and extends his original observations. Thermolysis on silica gel is intro- duced as an alternative technique for affecting rearrangements which appear to proceed via the acrylic ester (59). For example, stemmadenine-0-acetate was heated at 150° on a silica gel surface for 25 minutes; this treatment generated 0.15-0.2% (+)-vinca- difformine (75) (14,15-dihydrotabersonine). Similar treatment of (+)-allocatharanthine (60) yielded (+)-pseudocatharanthine (57)(4%) and optically pure (-)-tabersonine (56)(4%).

Kutney and co-workers have reported feeding experiments[79] bearing on the role of the acrylic ester (59) in indole alkaloid biosynthesis. 16,17-Dihydrosecodin-17-ol (63, R = OH) and secodine (76) were administered to *C. roseus via* the cotton wick technique. Only secodine showed significant incorporation into vindoline (6); three doubly-labeled forms of (76) were then fed to *C. roseus* and intact incorporations (0.03-0.07%) proven by degradation.

Finally, Scott has provided an account[80] of some fascinating experiments involving short-term incubation of labeled tryptophan with young (9-17 days) *C. roseus* seedlings. $2\text{-}^{14}\text{C-DL-Tryptophan}$ was taken up rapidly by the young plants and the radioactivity in various alkaloids was monitored vs. time. A new alkaloid of unknown structure appeared within five minutes of the start of the incubation period and this metabolite contained 35% of the total activity in the alkaloid fraction at the time. Within one hour, the unidentified compound contained virtually no radioactivity. Radioactive tabersonine (6) appeared rapidly and after nine hours, the level of radioactivity reached a maximum corresponding to a 30% incorporation of tryptophan into the alkaloid! Autoradiograms showed that tabersonine was metabolized almost as rapidly as it was formed and its activity fell from 12% of the total (9 hrs.) to 2% in three days. The radioactivity of geissoschizine (53) rose rapidly over a 60-90 minute period and gradually declined over the eight day period of the experiment. The interested reader is referred to the original paper for further details of these revealing experiments.

CHAPTER VI

TISSUE CULTURE STUDIES OF CATHARANTHUS ROSEUS

David P. Carew

College of Pharmacy, University of Iowa,
Iowa City, Iowa

INTRODUCTION

A simple definition of a plant tissue culture might be
that it is the successful growth of a mass of plant cells on
some type of nutrient medium. The mass of plant cells can be
part of an organ such as a root or stem or it may be simply a
mass of undifferentiated cells (callus). Most commonly, a plant
tissue culture refers to a culture of callus tissue growing on
a solid medium (static culture) or in a liquid medium (suspension culture). The history, techniques, and procedures for
culturing plant cells on artificial media have been described
in several publications.[1-6]

Tissue culture studies of *Catharanthus roseus* have been
either with crown gall tissue, which originated from an infection by the crown gall bacterium *Agrobacterium tumefaciens*
(Phytomonas tumefaciens), or with callus cultures which have
been initiated by placing portions of a normal *C. roseus* plant
on a medium containing growth regulators. For the purpose of
this chapter the author has chosen to discuss *C. roseus* tissue

culture in two parts, namely crown gall or tumor culture and
callus culture.

CROWN GALL TISSUE CULTURE

In 1941 Kunkel[7] reported that the normal periwinkle, *Catharanthus roseus* was found to be very susceptible to a viral infection known as "aster yellows". The virus, however, could be
inactivated if the infected plants were exposed to a suffi-
ciently high temperature for a two week period. Kunkel found
that the plant could survive this exposure and thereafter grow
as normal. Using the knowledge provided by Kunkel, Braun[8], who
was studying plant tumor inception, later theorized that *C.
roseus* might be inoculated not with a virus but with a bacterium
(*Phytomonas tumefaciens*) which in turn also could be killed
when the plant was exposed to heat. Thus a study of tumor production by the crown gall bacterium might be carried out since
the bacterium could produce the tumor and the latter could
continue to grow in the absence of the bacterium which would
have been destroyed by heat treatment. Braun's research
revealed that normal *Catharanthus* host cells could easily be
inoculated with the bacterium but when *Phytomonas tumefaciens*
was permitted a 36 to 48 hour incubation period prior to heat
treatment, only a small percentage of tumors formed. If the
incubation period was three days long then much greater tumor
formation resulted. Later, Braun showed that both slow and
fast growing tumors of *C. roseus* were capable of autonomous
development.[9]

White[10] isolated tissue from the crown gall tumors initiated by Braun and he successfully cultivated this tissue on a
solid medium for several months. He then grafted fragments
of these cultures into healthy, young *C. roseus* plants and
after a period of time tumors were produced. White then
proceeded to hypothesize as to the mode of tumor formation.

In discussing some of the implications of his work, he noted
that *C. roseus* belongs to the Apocynaceae which are noted
among herbaceous plants for producing a latex. Some of the
large cells of tissue cultures obtained with his grafting
experiments had a resemblance to latex cells. He thought that
it might be possible to use these tumors to study the mechanism
of latex formation. Plant tissue cultures have been used to
study latex formation.[11]

   The study of tumor initiation from crown gall infection of
C. roseus was continued by Braun[12] as he grew infected plants
at different temperatures to study the effect of temperature on
the physiological activity of host cells as they are altered to
form tumor cells.

   In 1947 de Ropp conducted experiments with normal and crown
gall tumor tissue of *Catharanthus*[13]. His studies were concerned
chiefly with determining the response of normal and tumor tissue
to synthetic growth hormones including indoleacetic acid, indole-
butyric acid and naphthaleneacetic acid. It has now been established that crown gall tumor cells do have the capacity to elicit
their own growth hormone.

   Braun,[14] and Braun and Wood[15] have discussed crown gall
tumor formation at length and have shown that with the crown
gall cell there is a series of quite distinct, but well-defined
systems of growth factor synthesis which progressively become
activated as the normal cell is altered to become a tumor cell.

Nutrition

   The growth of *Catharanthus* crown gall tumor tissue has
been accomplished on a number of different nutrient media. Most
formulations, however, represent variations of a few basic media.
Among the most commonly used basic media are those of White and
Braun,[16] and White.[17,18] These basic formulae are comprised of

several inorganic salts at macro and micro concentrations, certain members of the vitamin B complex, glycine, sucrose and agar.

In a comprehensive study of the nutrition of several plant
tissue cultures, Hildebrandt and Riker[19] found that while dextrose and sucrose are excellent carbon sources for *C. roseus*
crown gall cultures, galactose, maltose, lactose, cellobiose
and raffinose will also support growth. These workers also
found that the presence of pyruvic acid as well as propionic
and stearic acid can be beneficial for *Catharanthus* crown gall
tissue. Furthermore, these authors noted that among several
alcohols screened, butanol inhibited the growth of all cultures
in the study except *Catharanthus*. In later studies Hildebrandt
and Riker[20] again confirmed that dextrose, lactose and sucrose
are excellent carbon sources with an optimum concentration
being at 1 to 2 percent. In 1954 Hildebrandt *et al.*[21] studied
the growth of *Catharanthus* crown gall on media containing 16
organic acids in concentrations ranging from .015 to 4%. If no
other source of carbon was present in the medium, the tissue
did not grow when any of the 16 organic acids was added. When
2% sucrose was present, however, the tumor tissue tolerated a
wide range of concentrations of many of the organic acids.
While this tolerance was exhibited by the tissue there still
was not very much stimulation of growth with the exception of
that provided by oxalic, tartaric and pyruvic acids. Overall,
the *Catharanthus* tissue was more tolerant to the wide range
of compounds and concentrations than the other tissues in the
study. Braun[14,15] has done extensive work with plant tumors
and among his many reports was one in which he and Wood[22] compared the growth of normal *Catharanthus* cells and tumor cells
growing on White's basic medium. They found that normal *Catharanthus* cells grew very well on the White's basic medium provided that it was fortified with certain inorganic salts. They
also showed the importance of the presence of a certain level

of potassium ion for growth of tissue. According to Braun and
Wood one of the fundamental differences between a normal plant
cell and a tumor cell involves ion transport and membrane per-
meability.

The importance of meso inositol was recognized by Wood
and Braun[23] in a comparative investigation of normal and tumor
tissues of *Catharanthus*. It appears that meso inositol can
greatly facilitate the transport and utilization of certain
inorganic ions by *C. roseus* cultures. In the same study the
authors incorporated different levels of nitrate, phosphate
and potassium into a basic medium and they concluded that the
tumor cells possessed a more effective transport system than
the normal cells since the tumor cells grew better on the
basic medium. Boder *et al.* has conducted research with the
liquid culture of *C. roseus* crown gall and they noted that the
presence of vitamin K and chlorophyll in the medium were responsible for greatly stimulated growth.[24]

Growth Factors

Crown gall tissue normally supplies its own source of
auxin or growth regulator and an exogenous supply is not necessary. Normal tissue, however, if it is to be grown as callus,
will require some form of growth factor in the medium. A more
detailed consideration of the growth regulators used with
Catharanthus cultures will be found under the discussion of
normal tissue cultures (callus cultures).

Growth Conditions

Most of the studies of *Catharanthus* crown gall tissue have
utilized an agar medium. The pH is ordinarily adjusted prior
to sterilization of the medium and will usually be between 5.5
and 6.0. Most cultures have been grown in the dark or in diffuse light and at temperatures ranging from 25 to 28°C. The

containers for media may be test tubes, round glass vials, small
bottles or conical flasks.

Occurrence of Alkaloids

Tissue cultures of *C. roseus* crown gall are capable of
producing small amounts of alkaloids.[24] These have been found
both in the tissue and the medium and vindoline was the major
alkaloid produced. When vindoline hydrochloride was added to
a *Catharanthus* crown gall suspension, the tissue was able to
effect several modifications of the vindoline molecule.

CALLUS TISSUE CULTURE

This portion of the chapter will be concerned with callus
tissue cultures of *C. roseus* which have been derived by culturing normal tissue on a nutrient medium which includes certain growth regulators.

One of the earliest reports of the culture of normal *C.*
roseus tissue was by de Ropp[13] who worked with cambial tissue
from stems. This tissue, along with *C. roseus* crown gall tissue,
was grown on White's medium. When certain auxins were added
to the basic medium there was a distinct difference in response
by the normal and tumor tissue. The normal tissue was extremely
sensitive to the presence of auxin even at a concentration of
1 part per 100 million, while the tumor tissue remained unaffected unless the auxin concentration was very high. de Ropp
concluded that the lack of response of the tumor tissue to
auxin was because the tissue had already generated an excess of
some growth hormone, a property which was lacking in the normal
cells.

Wood and Braun[23] studied the growth of normal and crown
gall tissue and they also used a basic White's medium. It was
their conclusion that normal cells, of the type from which tumor
cells are derived, do not appear to synthesize any auxin for

unless such a substance was added to the basic medium there was
very limited growth. In a later paper these same workers reported that *Catharanthus* callus could be maintained without an
exogenous auxin supply if certain inorganic salts plus kinetin
and inositol were present.[22]

In 1962 Babcock and Carew[25] cultured the tissues of several
members of the Apocynaceae, including *C. roseus*. These workers
established continuing cultures from stem and seedlings of *C.
roseus* and they found that cultures could be best maintained
if they were transferred to fresh media about every three weeks.
A comparative study of a modified White's medium and that of
Wood and Braun was conducted by Harris *et al.*[26] with callus
cultures of *C. roseus* growing on solid media. Those workers
also reported the first successful growth of submerged cultures
of *C. roseus*. Callus cultures were initiated from *C. roseus*
leaf, stem and root tissue by Richter *et al.*[27] and the resulting
growth was analyzed for its alkaloid content.

The first report of prolonged growth of suspension cultures
of *C. roseus* was by Carew[28] and he determined the growth rates
of the tissue in media with and without growth substances. The
tissue was cultured in 500 ml conical flasks containing 100 ml
medium. An inoculum of about 1.3 Gm. per flask was used and
agitation was accomplished with a rotary shaker. The usual
growth cycle included about a three week lag period followed
by three weeks of active growth.

The most comprehensive investigation of *C. roseus* tissue
culture so far reported is by Patterson[29] and Carew.[30] These
authors studied growth of *C. roseus* in static and suspension
culture including growth in fermentors. In addition there was
a three month experiment dealing with the influence of selected
antibiotics on the growth of static cultures of *C. roseus*. A
detailed examination of tissues and media for alkaloids was
also a part of their research.

Nutrition

While several different media have been used for the culture of *Catharanthus* callus most workers have used the basic
White's medium[6,17] or that of Wood and Braun[23] or some modification of these. Since these media have been commonly referred
to throughout this chapter, the author has listed the ingredients
of the White's and the Wood and Braun medium as modified for use
in his laboratory for the culture of *C. roseus* (see Table I).
Iron in the chelated form has proven to be more satisfactory
than ferric sulfate or ferric citrate or tartrate. Inositol is
present in the Wood and Braun modification but not in White's.
When growth factors are used, a combination of coconut water
and 2,4-D seems best for static cultures while kinetin and NAA
have been used in suspension culture media. The reason for the
higher concentration of certain inorganic salts in the Wood and
Braun medium has already been discussed.

Harris *et al.*[26] have grown static cultures of *C. roseus* on
both the modified White's and the Wood and Braun medium and the
growth was favorable with each. However, they concluded that
the growth rate and general appearance and texture of the tissue
was somewhat better when the modified White's formula was used.
With reference to vitamins, *C. roseus* tissue grows well in both
static and suspension culture when just niacin, pyridoxine and
thiamine are present.

Growth Factors

A number of growth factors or regulators have been included
in culture media in order to stimulate callus growth of *C. roseus*.
Babcock and Carew[25] used 2,4-dichlorophenoxyacetic acid (2,4-D)
in several concentrations as well as benzothiazole-2-oxyacetic
acid (BTOA), indoleacetic acid (IAA) and naphthaleneacetic acid
(NAA). Stem callus grew well on a medium with 2,4-D but growth
decreased when the 2,4-D was substituted with BTOA. Callus

TABLE I.

Composition of Modified White's and Modified Wood and Braun Media

White's (mg/L)		Wood and Braun (mg/L)
360.	$MgSO_4$	360.
200.	$Ca(NO_3)_2$	200.
200.	Na_2SO_4	200.
80.	KCL	910.
65.	KNO_3	80.
16.5	NaH_2PO_4	316.5
–	$NaNO_3$	1800.
–	$(NH_4)_2SO_4$	790.
4.5	$MnSO_4$	4.5
1.5	$ZnSO_4$	1.5
1.5	H_3BO_3	1.5
0.75	KI	0.75
–	$H_2MoO_4 \cdot 5H_2O$	0.0018
–	$CuSO_4 \cdot 5H_2O$	0.0195
55.5	NaFeEDTA	55.5
0.5	Niacin	0.5
0.1	Pyridoxine HCL	0.1
0.1	Thiamine HCl	0.1
3.	Glycine	3.
–	Inositol	100.
2.%	Sucrose	2.%
15.%	Coconut Water	–
3.	2,4-D	–
–	Kinetin	0.5
–	NAA	1.
0.7%	Agar	0.7%

growth which was good on a medium with 2,4-D was sharply decreased when IAA was used in place of the 2,4-D. Coconut water has been used for many years to support the growth of plant tissue cultures and it has been frequently included in media in which *C. roseus* has been cultured. This substance still has not been completely defined and for this reason many investigators prefer not to use it, if possible. Tulecke *et al.* elucidated the amino acids present in coconut water[31] and a mixture of these amino acids has been used as a replacement for coconut water in media in which *C. roseus* has been cultured. However, the amino acid mixture proved unsatisfactory as such a replacement.[25]

Yeast extract and casein hydrolysate have also been commonly employed to promote the growth of callus tissue. These have been less active than other growth factors with *C. roseus* cultures.

The level of growth factor is always important and different tissues may be more, or less, sensitive to the same concentration. Furthermore, it is often necessary to use one concentration of growth factor to initiate callus formation and then a lesser amount to maintain callus growth. This situation was found to be true for *C. roseus* leaf callus cultures.[29] This also is true when one establishes a suspension culture using static culture tissue as inoculum. *C. roseus* tissue which normally grows well on solid media containing 3mg/L of 2,4-D will not survive if this same concentration is present in a liquid system.[3]

The addition of inositol to media for *C. roseus* callus culture was first reported by Wood and Braun and this compound has proven to be very beneficial. Concentrations from 100 to 200 mg/L seem to be optimum. Interestingly, inositol is one of the several sugar alcohols which have been isolated from coconut water.

The necessity of providing growth factors to tissue indefi-
nitely has sometimes been questioned. Suspension cultures of
C. roseus have been grown in our laboratory for more than three
years on a modified Wood and Braun medium, with and without NAA
and kinetin. While tissue on each medium grew well for a prolonged period, eventually there developed an irreversible
decrease in the growth rate of tissue on a medium containing
NAA and kinetin. This same medium, but without these latter
growth factors, was able to provide continued rapid growth of
C. roseus suspension cultures.[29]

Growth Conditions

The conditions which were briefly described for the growth
of *C. roseus* crown gall cultures are very similar to those
required for normal callus culture. Cultures are commonly grown
in the darkness or in diffuse light. An ideal temperature range
is from 26-28° C and the optimum pH of the medium is 5.7-5.8.
Static cultures may be grown on media contained in test tubes,
bottles or flasks. Suspension cultures are usually contained
in flasks which are plugged with cotton and placed on a reciprocal
or rotary shaker.

Fermentor Studies

C. roseus tissue has been grown in fermentors of two and
five liter capacity.[30] Growth has been quite satisfactory provided that contamination can be avoided. Highest yields of tissue have been obtained in a New Brunswick five liter fermentor
and these average from 4 to 6 grams per liter per day. The normal culture period is about 50 days and the tissue appears to
go through two growth cycles in that period. However, if a
shorter period of growth is used, say 25 to 30 days, the yields
are much less than one half of those obtained in a 50 day period.

Occurrence of Alkaloids

The presence of alkaloids in *C. roseus* static cultures was first reported by Babcock and Carew.[25] In later work Harris *et al.*[26] noted the presence of alkaloids in static cultures, as well as in suspension tissue and, in the nutrient medium. In 1965 Richter and co-workers established *C. roseus* leaf and stem callus cultures and following extraction of this tissue they found both vindoline and vindolinine.[27]

The major research concerning alkaloid formation and accumulation in callus tissues and media has been the work of Patterson.[29] Here tissues from static and suspension culture as well as the media were extracted using the procedure of Svoboda[33] with modifications by Loub *et al.*[34] Eight primary fractions were obtained from each tissue or sample of medium and subjected to thin-layer chromatographic analysis. In certain fractions it was difficult to separate the alkaloids and when this occurred the alkaloids were fractionated by subjecting the particular extract to adsorption chromatography using a deactivated alumina column. Extracts were chromatographed on thin-layer plates along with certain available known *Catharanthus* alkaloids. The results of a portion of this study are in Table II.

Among the many alkaloids detected by thin-layer chromatography were three substances which did not correspond to any previously reported compounds. These were designated as alkaloids A, B and C. Alkaloids A and B were present in the greatest quantity of any of the alkaloids found in tissue and media. While a number of alkaloids were detected, and several tentatively identified by thin-layer chromatography, none of the compounds detected were dimeric alkaloids. Also, neither vindoline nor catharanthine was found. These observations clearly indicate that differences exist between the synthetic capability of the whole plant and that of isolated callus tissue. The over-all pattern of alkaloid distribution in tissue and media was similar.

TABLE II.

Alkaloids in Extracts of *C. roseus* Tissue and Liquid Media

Alkaloid	Source[a]
Alkaloid A	1,2,3,4
Alkaloid B	1,2,3,4
Lochneridine	1,2,3,4
Cavincidine	1,2,3,4
Sitsirikine	1,2,3,4
Akuammicine	1,2,3
Cathalanceine	1,4
Alkaloid C	2,3
Dihydrositsirikine	2,3
Ajmalicine or Cavincine	3,4
Cathindine	1
Mitraphylline	2
Serpentine or Alstonine	4
Perosine or Perivine	4
Lanceine	4

[a] 1= Modified Wood-Braun liquid medium with kinetin and NAA

2= Modified Wood-Braun liquid medium without kinetin and NAA

3= Callus tissue initiated from *C. roseus* seedlings

4= Callus tissue initiated from *C. roseus* leaves

Alkaloids A and B as well as lochneridine, sitsirikine and cavincidine were present in all tissues and media while others were found only in tissue or media. There was no appreciably

different pattern in alkaloid content of the leaf callus as compared to the seedling callus or media. Of interest was the fact
that yohimbine and cathalanceine or lanceine appeared to be
present in the cultures while these have not been reported to
be found in the normal *C. roseus* plant. Also, cavincine and
akuammicine were identified in leaf cultures although these
supposedly only occur in the roots of the normal plant.[32]

Miscellaneous Study

   Microbial contamination is an ever-present problem with
plant tissue cultivation and the inclusion of an antibiotic
in the culture medium might be one way of controlling contamination. However, the antibiotic could be toxic to the tissue
and therefore its safety must be established prior to its use.
Patterson and Carew[29] conducted a limited study utilizing *C.
roseus* tissue cultures and four different antibiotics, each
of which has a somewhat different spectrum. The antibiotics
studied were bacitracin, streptomycin, griseofulvin and
oxytetracycline and these were incorporated into agar medium
normally used to support the growth of *C. roseus* callus. The
medium was inoculated with *C. roseus* callus and growth allowed
to proceed for 30 days. At that time any growth increase was
determined and some of the tissue was inoculated on fresh media.
Data from three 30 day growth periods was obtained. The results
showed that bacitracin was essentially non-toxic to the tissue
while a level above 2.5 mg/L of streptomycin was quite toxic.
Griseofulvin was non-toxic at 2.5 mg/L and oxytetracycline did
not exhibit toxicity until a concentration above 10 mg/L was
reached. With the exception of streptomycin any of these
antibiotics might be considered for use as protective agents
against microbial contamination of *C. roseus* tissue cultures.

TISSUE CULTURE STUDIES OF C. ROSEUS 07

ACKNOWLEDGEMENTS

The author is grateful to Norman R. Farnsworth of the
University of Illinois and Gordon H. Svoboda of Eli Lilly and
Company for providing known alkaloids used in a portion of his
research. Part of the research reported here was supported by
grant HE 05290 from the National Institutes of Health.

REFERENCES

1. R.J. Gautheret, La Culture des tissus Végétaux, Masson et
   Cie, Paris (1959).

2. A.C. Hildebrandt, Moderne Methoden der Pflanzenanalyse,
   5 Band, Springer-Verlag, pp. 382-421 (1962).

3. D.P. Carew and E.J. Staba, Lloydia, 28, 1 (1965).

4. H.E. Street, J.U.S. Cancer Inst., 19, 467 (1957).

5. H.E. Street and G.G. Henshaw, Symp. Soc. Exptl. Biol., 17,
   234 (1963).

6. P.R. White, The Cultivation of Animal and Plant Cells, 2nd
   ed. The Ronald Press, New York (1963).

7. L.O. Kunkel, Am. J. Botany, 28, 761 (1941).

8. A.C. Braun, Am. J. Botany, 30, 674 (1943).

9. A.C. Braun, Phytopathology, 41, 963 (1951).

10. P.R. White, Am. J. Botany, 32, 237 (1945).

11. B. Arreguin and J. Bonner, Arch. Biochem. 26, 178 (1950).

12. A.C. Braun, Am. J. Botany, 34, 234 (1947).

13. R.S. de Ropp, Am. J. Botany, 34, 53 (1947).

14. A.C. Braun, Proc. Natl. Acad. Sci., 44, 344 (1958).

15. A.C. Braun and H.N. Wood, Advan. Cancer Res., 6, 81 (1961).

16. P.R. White and A.C. Braun, Cancer Res. 2, 597 (1942).

17. P.R. White, Ann. Rev. Biochem., 11, 615 (1942).

18. P.R. White, A Handbook of Plant Tissue Culture. The Jaques
     Cattell Press, Lancaster, Penn., 277 pp. (1943).

19. A.C. Hildebrandt and A.J. Riker, Am. J. Botany, 36, 74
     (1949).

20. A.C. Hildebrandt and A.J. Riker, Am. J. Botany, 40, 66
     (1953).

21. A.C. Hildebrandt, A.J. Riker and J.L. Watertor, Phyto-
     pathology, 44, 422 (1954).

22. A.C. Braun and H.N. Wood, Proc. Natl. Acad. Sci., 48, 1776
     (1962).

23. H.N. Wood and A.C. Braun, Proc. Natl. Acad. Sci., 47, 1907
     (1961).

24. G.B. Boder, M. Gorman, I.S. Johnson and P.J. Simpson,
     Lloydia, 27, 328 (1964).

25. P.A. Babcock and D.P. Carew, Lloydia, 25, 209 (1962).

26. A.L. Harris, H.B. Nylund and D.P. Carew, Lloydia, 27, 322
     (1964).

27. I. Richter, K. Stolle, D. Gröger and K. Mothes, Natur-
     wissenschaften, 11, 305 (1965).

28. D.P. Carew, J. Pharm. Sci., 55, 1153 (1966).

29. B.D. Patterson, Ph.D. Dissertation, University of Iowa,
     (1968)

30. B.D. Patterson and D.P. Carew, Lloydia, 32, 131 (1969).

31. W.L. Tulecke, L. Weinstein, A. Rutner and H. Laurencot, Jr.,
     Contrib. Boyce Thompson Inst., 21, 115 (1961).

32. G.H. Svoboda, A.T. Oliver and D.R. Bedwell, Lloydia, 26,
     141 (1963).

33. G.H. Svoboda, N. Neuss and M. Gorman, J. Pharm. Sci., 48,
     659 (1959).

34. W.D. Loub, N.R. Farnsworth, R.N. Blomster and W.W. Brown,
     Lloydia, 27, 470 (1964).

CHAPTER VII

BIOCHEMISTRY OF DIMERIC CATHARANTHUS ALKALOIDS

William A. Creasey

Departments of Internal Medicine and Pharmacology
Yale University, School of Medicine
New Haven, Connecticut 06510

INTRODUCTION

The periwinkle plant *Catharanthus roseus* G. Don, which has
commonly but incorrectly been called *Vinca rosea* Linn., has long
enjoyed a reputation in popular medicine throughout the world
for the treatment of such diverse conditions as hemorrhage,
scurvy, diabetes, wounds, and toothache. Attempts to investigate the reputed hypoglycemic effects of the plant, by groups
at the Collip Laboratories, University of Western Ontario, and
at the Lilly Research Laboratories, Indianapolis, led to the
isolation of a group of four dimeric indole alkaloids: vinblastine (vincaleukoblastine), vincristine (leurocristine), vinleurosine (leurosine) and vinrosidine (leurosidine).[1-5] These compounds, however, instead of depressing blood glucose levels,
caused a profound inhibition of cellular proliferation. Much
of the detail of these early studies has been reviewed elsewhere.[6]
Suffice to say that these chance findings gave to clinical cancer
chemotherapy vinblastine and vincristine, two of the most valuable
agents now available. A result of this mode of discovery was that

209

in contrast to most other chemotherapeutic agents, which were
developed with the specific aim of causing definitive metabolic
lesions in susceptible neoplastic cells, the biochemical pharmacology of the *Catharanthus* alkaloids remained largely unknown
long after their clinical usefulness had been well established.
The arrest of cell division at metaphase with the formation of
C-mitoses, an action characteristic of the group,[7-9] suggested
that they might act like colchicine, but the mechanism of action
of the latter was itself equally obscure. Another separate indication of a possible mode of action was given by the finding
of a number of antagonists of the new antimitotics; these included
aspartic, glutamic and α-ketoglutaric acids, arginine, citrul-
line, ornithine, tryptophan and coenzyme A.[10-12] In general,
however, the *Catharanthus* alkaloids appeared disconcertingly
inactive when tested against a variety of biochemical parameters,
unless the concentration of drug were raised to very high, clearly unphysiological levels.[13] Nevertheless, the continuing demonstration of their clinical effectiveness provided a powerful
stimulus to several groups of workers, including our own, to
attempt to elucidate their mechanism of action. Now, more than
ten years after the initial demonstration of oncolytic activity
in the *Catharanthus* extracts, it is at last becoming possible
to speak with some degree of confidence about the mechanisms by
which these agents exert their effects. This has been accomplished through a variety of biochemical and cytological approaches. It is my intention in this chapter to review those
salient steps along the pathway of our increasing knowledge
that have made it possible to reach the present stage. At the
same time, it is hoped to give a picture of how these alkaloids are playing a part in fundamental studies of the mechanism of cellular growth and function.

NUCLEIC ACID BIOSYNTHESIS

The major role played by nucleic acids in cell growth and proliferation has led to an emphasis upon the development of inhibitors of nucleic acid biosynthesis as potential chemotherapeutic agents. It may be questioned whether concentration on this particular metabolic area is entirely justified in view, for example, of the vital importance of cell surface phenomena where lipids play a large role, but a practical consequence is that potential new agents are invariably screened as possible inhibitors of the biosynthesis of RNA and DNA.

Examples of Inhibitory Effects

The first such screening of the *Catharanthus* alkaloids failed to show any effect of either vinblastine or vincristine upon the incorporation of formate-^{14}C or glycine-2-^{14}C into RNA and DNA by sarcoma 180 ascites cells *in vitro*.[13] Similar negative findings were reported by Beer[14] with respect to nucleic acid synthesis in regenerating liver and Ehrlich ascites carcinoma. Our initial approaches to this problem produced essentially analogous results. Levels of vinblastine in excess of 1×10^{-4}M were necessary before significant depression of the incorporation of thymidine-^{3}H into DNA could be demonstrated in Ehrlich ascites cells *in vitro*; uptake of uridine-^{3}H into RNA appeared to be even less sensitive. However, when attention was turned to the effects of treatment of tumor-bearing mice, a different picture emerged. With the Ehrlich ascites tumor, single doses (2 mg/kg) of vinblastine produced a time-dependent reduction of the incorporation of uridine-^{3}H into a total phenol-extractable RNA. About 6 hours were required to achieve maximum depression, and full recovery occurred in 18 to 24 hours. Extraction of RNA from the cells with phenol and

fractionation on columns of methylated serum albumin using
linear gradients of sodium chloride, showed that the synthesis
of soluble or transfer RNA was most sensitive to the effects
of the alkaloid. The synthesis of DNA as measured by incorporation of thymidine;[3]H did not appear to be much affected.[15]
Love[16] has reported a similar specific action of colchicine on
the ribonucleoproteins of Ehrlich cells, but this specificity,
like that found in our laboratory for vinblastine, may not be a
general phenomenon. In sarcoma 180, a tumor generally considered
to be similar to the Ehrlich carcinoma, vinblastine markedly
inhibits the synthesis of both RNA and DNA; this is clearly seen
in Figure 1. Furthermore, we have found that the incorporation
of thymidine-^{3}H into DNA by Ehrlich ascites cells *in situ* is
approximately as sensitive to treatment with colchicine or vincristine as is RNA synthesis.[17] Recently, Wagner and Roizman[18]
have reinvestigated the apparent selective sensitivity of transfer RNA to vinblastine using HEp-2 cells in culture. They
found that the synthesis of transfer RNA itself is if anything
less sensitive than that of ribosomal RNA, but that incorporation of radioactivity into another low molecular weight material
was severely depressed by vinblastine. This unknown component
could only be resolved from transfer RNA by acrylamide gel elec-

trophoresis. Thus, the synthesis of transfer RNA, isolated by
column chromatography or sucrose density gradient ultracentri-
fugation, might appear to be selectively inhibited because of
the effect of vinblastine on this contaminant. Our studies were
later extended to vincristine, and comparisons made with colchicine.[19] Although both of these drugs appeared to show some
degree of selectivity for the soluble RNA fraction, there were
more pronounced effects on rapidly-labeling interphase and ribosomal RNA in the treated Ehrlich cells than had been encountered
with vinblastine. It was of great interest that pretreatment
of the tumor-bearing mice with massive doses (0.9 to 2.7 g/kg)

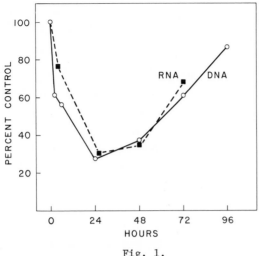

Fig. 1.

Effect of a single dose of vinblastine (2 mg/kg) at time
zero upon the incorporation of deoxycytidine-^{3}H into DNA and
of uridine-^{3}H into total RNA of sarcoma 180 ascites cells *in
situ*. Values were determined from the amount of labeled precursor incorporation per 10^8 cells after a one-hour period for
metabolic utilization.

of glutamic acid decreased the inhibition of RNA synthesis produced by the alkaloids; this observation recalls the antagonism
to the antitumor and anti-mitotic effects noted when the agents
were first isolated.[10-12] The presence of a significant time
factor in the development of the depression of RNA synthesis
seen *in vivo*, led us to reexamine the system used *in vitro*, when
concentrations of vinblastine as high as 6 x 10^{-4}M had no effect
on the incorporation of uridine-^{3}H.[15] Using a prolonged (1
hour) preincubation of Ehrlich cells with drug, it was found
that the uptake of uridine-^{3}H into total RNA, measured as cold
acid-insoluble, alkali-hydrolyzable radioactivity, was depressed
significantly; vinblastine was the most active alkaloid giving

a 58 percent inhibition at 2×10^{-4}M.[20]   Further, the protection
afforded by glutamic acid *in vivo* could also be demonstrated *in
vitro*, when preincubations were carried out with a 30 - to 60 -
fold excess of the amino acid.

Meanwhile many other workers were studying the interaction
of the *Catharanthus* alkaloids with nucleic acid biosynthesis.
The Canadian group, which had pioneered in the introduction of
these agents, reported on their experiments with rat thymus,
bone marrow and chloroleukemic cell suspensions.[21,22] (In rat
thymus cells, vinblastine was an active inhibitor of both RNA
and DNA synthesis, measured by incorporation of formate-[14]C
and glycine-2-^{14}C; its activity was closely approached by vinleurosine, but exceeded that of vincristine and vinrosidine.)
The inhibitory action of vinblastine was more marked on the
specific activity of the purine bases in DNA than on the pyri-
midines.) Van Lancker and his coworkers examined rat spleen and
bone marrow[23] and regenerating liver.[24]   In these tissues, incorporation of thymidine-^{3}H into DNA was markedly depressed by
vinblastine, the minimum specific activity occurring during the
12 to 48 hour period in the spleen and marrow. Incorporation of
cytidine-^{3}H and orotic acid-^{14}C into RNA was essentially unchanged, except for a small reduction in the activity of nuclear
RNA. A contrary finding was reported for rat thymus and lymphosarcoma P1798 in which there was a decrease in RNA content without definitive changes in DNA following treatment with vinblastine; at the same time there was a fall in ribonuclease
activity.[25]   In rats bearing the Walker tumor, Desjardins *et al.*[26] compared the effects of uracil mustard, actinomycin D and vinblastine upon nucleic acid synthesis. They found that vin-
blastine produced a relatively greater degree of inhibition of
the uptake of thymidine-^{3}H into tumor nucleolar, as compared
with non-nucleolar DNA. Even DNA synthesis in protozoa appears
to be sensitive to the *Catharanthus* alkaloids, as indicated by

[handwritten marginal note: Action on purine base rather than pyrimidine base!]

the profound depression in the incorporation of uridine-^{3}H and
thymidine-^{3}H reported for *Tetrahymena pyriformis* in the presence
of vincristine;RNA synthesis, although inhibited, was much less
sensitive to this agent.[27] These inhibitory effects are re-
versed by riboflavin, flavin mononucleotide, and flavin adenine
dinucleotide[27] apparently because of a photochemical inactivation of the alkaloids that is catalyzed by the flavins.[28] Cline[29]
has reported that in human acute lymphocytic leukemic leukocytes exposed to vincristine at 7×10^{-5}M, incorporation of
uridine-^{3}H into RNA was inhibited; with exposure times of up
to 24 hours an effect could be demonstrated at concentrations
as low as 7×10^{-9}M. In studies carried out in our laboratories,
leukocytes from patients with chronic granulocytic leukemia and
leukosarcoma incorporated radioactivity from cytidine-^{3}H into
both RNA and DNA; significant inhibition of this uptake was seen
in the presence of vincristine at concentrations of 0.5 and 1.0
$\times 10^{-4}$M.[17] Normal human granulocytes do not show any sensitivity
to vincristine, in terms of RNA synthesis, at these concentrations.[30] A final aspect of the interaction of *Catharanthus*
alkaloids with nucleic acid synthesis relates to neurological
toxicity. The latter is a feature associated more with vincristine than vinblastine. Agustin and Creasey[31] have found that
in mice treated with vincristine by daily intraperitoneal injection at dosages of 0.2 to 0.5 mg/kg, the incorporation of
uridine-^{3}H, administered by the intracranial route, into brain
RNA is greatly reduced. This effect, seen after only three
days of treatment, preceeds the appearance of overt neurologic
toxicity. Vinblastine produces depression of RNA synthesis at
a much later time, when the animals are moribund.

In Table I a summary is given of the reported examples of
inhibition of nucleic acid biosynthesis by the *Catharanthus* alkaloids.

TABLE I

Reports of Inhibitory Effects of the
Catharanthus Alkaloids on the Synthesis of Nucleic Acids

System	Drug	Nucleic Acid	Inhibitory Effect	Ref.
Ehrlich ascites	VLB	RNA DNA	*In vivo* *In vitro*	15
Ehrlich ascites	VCR	DNA	*In vivo*	17
HEp-2 cells	VLB	RNA	In culture	18
Ehrlich ascites	VLB,VCR	RNA	*In vivo*	19
Ehrlich ascites	VLB,VCR	RNA	*In vitro*	20
Rat thymus, bone marrow,chloroleukemia	VLB,VCR, VLR,VRD	RNA,DNA	*In vitro*,selective for purines	21,22
Rat spleen, bone marrow,regenerating liver	VLB	RNA,DNA	*In vivo*, RNA effect only on nuclear fraction	23,24
Rat thymus,P1798 cells	VLB	RNA	Fall in RNA content, no change in DNA	25
Rat Walker tumor	VLB	DNA	*In vivo*,especially nucleolar DNA	26
Tetrahymena pyriformis	VCR	DNA,RNA	RNA affected less than DNA, flavin reversal	27
Human leukemic cells	VCR	RNA	*In vitro*	29
Mouse brain	VCR	RNA	*In vivo*	31
Rat thymus	VLB	DNA,RNA	*In vivo*	32
Sarcoma 180	VLR	DNA,RNA	*In vivo & in vitro*	33

Abbreviations: VLB, vinblastine; VCR, vincristine; VLR vinleurosine; VRD, vinrosidine

Mechanism of Inhibition

An inhibitory effect on a biochemical pathway may arise either by direct action of the pharmacological agent on the system that is being measured, or by some indirect or abscopal action, as for example by influencing hormone production. In the case of the *Catharanthus* alkaloids, most of the evidence, using both *in vivo* and *in vitro* systems, points towards direct effect on nucleic acid synthesis. Nevertheless there are some factors that may contribute indirectly to the inhibitory effects that are observed. The release of adrenal steroids, provoked by vinblastine, might be expected to influence rates of nucleic acid synthesis, and although adrenalectomy did not appear to modify the degree of inhibition of nucleoside incorporation that was observed in the rat,[32] the possibility remains that under certain conditions steroid release might contribute to the overall inhibitory effect. Another possible source of in- direct action is the reported inhibitory effect of the alkaloids on the uptake of nucleosides by cells. In our laboratory, we have found that the uptake of uridine-^{3}H by human leukosarcoma cells is inhibited by vinblastine at somewhat higher levels than those that inhibit its incorporation into RNA.[17] The up- take of uridine-^{3}H by sarcoma 180 cells *in vitro,* and its sub- sequent incorporation into RNA were equally inhibited by vin- leurosine. Uptake of deoxycytidine-^{3}H was not affected, how- ever, despite a reduction in its incorporation into DNA.[33] Much earlier, McGeer and McGeer[34] reported a reduction in urinary excretion of 4(5)-amino-5(4)imidazole-carboxamide after treat- ment with vinblastine. This observation, suggestive of a block on the pathway of purine biosynthesis, may relate to reports of a selective inhibition of purine nucleotide incorporation into nucleic acids[22] and of apparent changes in the size of the cel- lular pool of glycine.[22,35] All of these effects could con-

tribute to the well-documented inhibition of nucleic acid biosynthesis.

A number of workers have studied the interaction of antimitotic alkaloids with polymerase enzyme systems in order to
determine whether there is a truly direct action on the synthetic process. Cline[36] reported that in a cell-free system
from normal rat spleen, vincristine caused a moderate degree
of inhibition of DNA-dependent RNA polymerase. Creasey,[37] using
a crude aggregate RNA polymerase preparation from mouse brain
as well as a purified bacterial RNA polymerase, was able to
demonstrate inhibition of the enzyme by relatively high levels
of vincristine. Another recent report[38] described inhibition
of DNA-dependent RNA polymerase from Ehrlich ascites carcinoma
cells by both vincristine and vinblastine. Nothing is yet
known of the possible molecular basis for this enzyme effect,
but it is tempting to draw an analogy with two other antimitotic
agents, colchicine, which on the basis of changes in optical
rotation, has been reported to bind to DNA,[39] and griseofulvin,
which binds to the RNA of sensitive fungi;[40] both these compounds are inhibitors of nucleic acid synthesis.

PROTEIN BIOSYNTHESIS

Disturbances in the synthesis of proteins would be expected
to occur ultimately as a result of inhibition of the biosynthesis of nucleic acids, in view of the central role played by the
latter in the fabrication of protein molecules. Many of those
who have produced evidence for inhibitory effects of the *Catharanthus* alkaloids on nucleic acid formation have also reported
similar findings for the incorporation of amino acids into protein. This has been true of our own studies with the Ehrlich
tumor,[19,20] in which the degree of inhibition of protein synthesis by vinblastine and vincristine was comparable in magnitude
to the effects on RNA synthesis. However, there was evidence

for a certain degree of selectivity, in that there was a depression in the uptake of valine-^{14}C and glutamic acid-^{14}C into protein, in the presence of vinblastine, while the incorporation of glycine-^{14}C was unaffected. In the case of glutamic acid, the actual transport of the amino acid into the cells was also depressed by vinblastine, a phenomenon not seen when uptake of glycine was studied. This has recently been examined in our laboratory with regard to normal human leukocytes, which are about 80 percent polymorphonuclear cells. In these experiments the incorporation of valine-^{14}C into protein was depressed by vincristine at concentrations ($<10^{-4}$M) that did not inhibit RNA synthesis, suggesting that in this system the inhibitory effect of the alkaloid is not mediated through RNA. Total cellular uptake of glutamic acid-^{14}C was inhibited by vinblastine; competitive kinetics were established for the mutually antagonistic uptake of these two compounds.[30] In leukemic lymphocytes, protein synthesis has also been described as the parameter most sensitive to vinblastine and vincristine.[38] On the other hand, in human leukemic leukocytes Cline[29] has found both processes equally sensitive to vincristine and showing a similar time course for development of inhibition; the same situation has been described in *Tetrahymena pyriformis*.[27]

RESPIRATION AND CARBOHYDRATE METABOLISM

The concept developed by Warburg[41] that many neoplastic cells may have a greater dependence on glycolysis for energy production than those of normal tissues, with correspondingly increased formation of lactate, has long provided a challenge to oncologists to develop chemotherapeutic agents effective in this metabolic area. Although little success has attended these efforts, many successful antineoplastic drugs, including the *Catharanthus* alkaloids, have been screened for such activity. Both vinblastine and vincristine modified respiration and lactic acid

production by cells *in vitro*.[13,14,42] Inhibition of respiration
required higher concentrations of drug than did suppression of
the Pasteur effect. In Jensen's sarcoma, Yoshida ascites,
hepatoma AH 130, and jejunal mucosa cells, vinblastine inhibited
anerobic glycolysis. A temporary abolition of this effect by
nicotinamide suggested that the lesion involved the pyridine
nucleotides.[43] It is, however, questionable whether these effects represent a fundamental mechanism of cytotoxicity or mitotic arrest, because similar disturbances in the biochemistry
of respiration have been reported for other, unrelated chemotherapeutic agents having different biological effects.[44] In any
event the evidence for the existence of high energy requirements
for mitosis is conflicting. Bullough[45] obtained data suggesting
that a critical energy reserve is needed for successful com-
pletion of mitosis in mammary carcinomas. In sea urchin eggs
there was a correlation between ATP content and mitotic activity,
but in pea root tips a profound reduction in the level of this
nucleotide did not affect mitosis.[46]

LIPID METABOLISM

     The metabolism of lipids is an area to which little atten-
tion has been paid in the design and development of cancer chemotherapeutic agents. Nevertheless, there is good evidence that
several antineoplastic drugs, including the *Catharanthus* alkaloids, exert effects on lipid metabolism. The neurologic toxicity so characteristic of vincristine seems to be associated
with an axonal degeneration, that includes extensive demyelina-
tion, in monkeys,[47] rodents and man.[48,49] Phospholipids are, of
course, major components of the myelin sheath. A less direct
indication of a lipid involvement comes from reports of cross-
resistance between vincristine and the phthalanilides,[50] which
are known to form complexes with lipid components.[51] In the
studies of the turnover of radioactive phosphate in rat gastroc-

nemius muscle, it has been found that vincristine reduces phos-
pholipid synthesis without affecting the metabolism of organic
acid-soluble phosphates.[52] Recently, we have examined lipid
synthesis in sarcoma 180 ascites cells. Vinleurosine exerted
an inhibitory effect both *in vivo* and *in vitro*, with marked
specificity for phospholipids. Indeed, synthesis of the latter
was more sensitive to the drug than was the synthesis of nucleic
acids or protein.[33] Vincristine displayed a similar specific
action upon the incorporation of acetate-^{14}C into phospholipids;
an example of these experiments appears in Figure 2.

MISCELLANEOUS BIOCHEMICAL EFFECTS

     Several biochemical effects of the *Catharanthus* alkaloids
have been described which do not fall into the major metabolic
areas considered thus far. An early report[53] described a fall
in the level of coenzyme A in rat liver after administration of
the total alkaloid fraction from the periwinkle plant. This
decrease was prevented by pantethine, but not pantothenate or
pantothenylcysteine, and the authors concluded that the decar-
boxylation of pantothenylcysteine was inhibited by the alkaloids;
such blockade could obviously have repercussions on both lipid
and carbohydrate metabolism. Bariety and Gadjos[54] studied the
effects of cyclophosphamide, 4-amino-10-methyl folic acid and
vinblastine on rabbits. All three compounds produced a significant elevation of free iron levels in the blood with recovery
to normal concentrations in 2 weeks; incorporation of iron into
protoporphyrin was unchanged, suggesting that a hemolytic process
might have been responsible. In addition, vinblastine produced
an elevation in blood copper content not seen with the other
agents. From a consideration of structural features, notably
the-$COOCH_3$ groups at positions 3 and 18', Moncrief and Heller[55]
proposed that the alkaloids may function as acylating agents;
model chemical systems were used to explore this possibility.

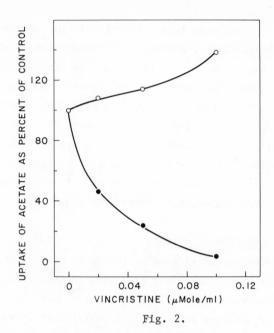

Fig. 2.

Effect of vincristine on the incorporation of sodium acetate-2-^{14}C into the neutral lipids (o——o) and phospholipids (●——●) of sarcoma 180 cells *in vitro*.

The *Catharanthus* alkaloids are indole derivatives, and since reserpine, which releases catecholamines and histamine from storage sites, also has a polycyclic structure containing an indole nucleus, it is logical to see if they produce a similar effect. In addition, vinblastine is known to profoundly alter the morphology and behavior of rat mast cells which are rich in histamine.[56] However, when vinblastine was tested for an effect on histamine release by mast cells it was found that rather than stimulating this process, it actually inhibited the release produced by other agents.[57]

STUDIES WITH CYTOLOGICAL CORRELATIONS

While considerable progress has been made by purely biochemical approaches, it is from an area of mainly cytological
study that much light has been shed, not merely on the mechanism
of action of this interesting group of agents, but also on the
whole question of cell growth and division. Much of this work
has involved parallel studies with colchicine and griseofulvin,
compounds that are able to produce very similar biological effects to the *Catharanthus* alkaloids in terms of their action
on the mitotic spindle.[7-9,58] Metaphase arrest is only one
aspect of this picture. The chromosomes may assume a super-
contracted form and be disoriented within the cell, while the
mitotic spindle and the microtubules associated with centrioles
and centromeres undergo a reversible dissolution.[58-61] In
L-strain fibroblasts and human leukocytes, exposure to vinblastine and vincristine gives rise to uniaxial birefringent
crystals, with sides up to 8μ long, which in electron micro-
graphs appear to be regular aggregates of microtubular bodies
sharing their walls in common in a hexagonal arrangement.[62]
Similar crystals have been observed in neurons from a patient
treated with vincristine intrathecally,[63] and in starfish
oöcytes exposed to vinblastine.[64] White has also reported the
disappearance of microtubules and the formation of inclusions
in human platelets treated with colchicine, vinblastine and
vincristine.[65] These cytological changes, all involving interactions with microtubules or organelles derived from them, have
been suggested as a basic mechanism underlying the anti-inflam-
matory and antimitotic effects of the metaphase arresters.[66,67]
Such studies find biochemical correlations in evidence of intra-
cellular binding of these drugs. Most of this work has involved
colchicine, the first metaphase-arresting agent to be available

Chromosome
be in
Disoriented
within
the
cell

?

with isotopic label. Perhaps the most complete of these studies
is that undertaken by Taylor and his coworkers, reported in a
series of papers.[68-72] Tritiated colchicine was bound by non-
covalent linkage to a protein that occurred in the soluble fraction of the cell after homogenization; the protein had a sedimen-
tation constant of 6S by zone centrifugation. Highest binding
activity was found in dividing cells, mitotic apparatus, cilia,
sperm tails and brain tissue, and there was a correlation between
degree of binding and the presence of microtubules in the tissue.
The active protein was extracted from the sperm tails of the sea-
urchin *Strongylocentrotus purpuratus*. It was concluded that
the binding protein is actually a subunit of the microtubules.
Wilson and Friedkin[73] have investigated the binding of colchicine
by a similar protein, of molecular weight 105,000, from grass-
hopper embryos. Binding was a time- and temperature-dependent
process that was inhibited by the antimitotics podophyllotoxin
and picropodophyllotoxin, and stimulated by vinblastine. Urea
destroyed the complex. In our laboratory we have studied the
binding of tritiated colchicine, vinblastine, and griseofulvin
by sarcoma 180 ascites cells, by human leukocytes, and by rab-
bit peritoneal exudate cells.[30,37,74-76] The findings were
in most respects similar to those of other workers, but additional attention was paid to the interaction of different antimitotic agents and other compounds. Data of this nature appear
in Table II, which represents the amount of binding, assayed by
equilibrium dialysis, between colchicine-^{3}H or vinblastine-^{3}H
and the 100,000 x g supernatant from homogenates of sarcoma 180
cells. The stimulations observed may in actual fact result
from stabilization of the complexes. Another metaphase arrester,
podophyllic acid ethyl hydrazide, has also been reported to bind
to proteins in mouse mammary tumors.[77]

The most logical explanation of the results of these investigations is that there are two types of binding sites on

TABLE II

Effect of Various Compounds on the Binding of Antimitotic
Agents by the Soluble Fraction of Sarcoma 180 Ascites Cells

Compound	Concentration (M)	Bound Alkaloid % Control Colchicine	Vinblastine
None		100	100
Colchicine	10^{-4}		184
Griseofulvin	10^{-4}	32	187
p-Fluoro-phenylalanine	10^{-4}	61	180
Vinblastine	10^{-4}	207	–
Vincristine	10^{-4}	218	40
Vinglycinate	10^{-4}	128	89
Glutamate	10^{-3}	100	93
NaCl	1.0	113	13
Urea	1.0	14	136

microtubular precursor material, one to which colchicine, griseofulvin and podophyllotoxin bind, and the other for which the
Catharanthus alkaloids, and also guanosine triphosphate,[77] have
an affinity. The Catharanthus site may involve an ionic type
of binding, as is suggested by the salt effect, whereas the colchicine site may depend upon hydrophobic interactions or hydrogen
binding and thus be sensitive to urea. The fact that only the
Catharanthus alkaloids, and not colchicine or griseofulvin produce microtubule crystals,[62] is compatible with this view of two
different kinds of interacting sites. An apparent concentration
of grains over the crystals in radioautographs of starfish oöcytes

exposed to tritiated vinblastine,[64] may indicate that the drug
itself forms an integral part of the microtubular aggregate. The
spacing of the tubules within the crystals[62,79] is such as to
suggest that changes in configuration of the tubules have taken
place.[62] This could conceivably be due to the substitution of
vinblastine or vincristine for guanosine triphosphate which may
normally occupy one of the binding sites. It is interesting
that addition of relatively high concentrations of vinblastine
to cell supernatant fractions causes precipitation of a pseudo-
crystalline aggregate that resembles the crystals formed *in vivo*,[79-83]
although it may contain other proteins beside tubule subunits.[84]

The importance of such studies lies in what it tells us
regarding the role of microtubules in normal cellular processes.
As components of the mitotic apparatus,[85] of animal and plant
flagella,[86,87] human leukocytes,[88] cellular microspikes and
cortical cytoplasm,[89] and of neuronal processes,[90] the microtubules are in a position to play diverse and vital roles in
cellular physiology. The availability of agents that are able
to act on the microtubules provides a powerful tool for evaluating the individual roles of these ubiquitous cellular components. Already, the variety of biological effects of antimitotic agents – mitotic arrest, cytotoxicity, interference with
phagocytosis, interference with histamine release, relief of
acute gouty episodes – provides an indication of the magnitude
of these roles.

METABOLISM AND DISTRIBUTION

Studies of the metabolism and distribution of the *Catharanthus* alkaloids have mainly been carried out with vinblastine.
Early data obtained by biological assay of serum from rats
treated with vinblastine at a dosage of 5 mg per kilogram indicated that drug levels were undetectable later than 5 minutes

after intraveneous injection.[91] Using a tissue-culture method
for assay, serum levels of vincristine of 0.3 to 1 µg/ml were
found in dogs and monkeys 5 minutes after injection (1 mg/kg),
decreasing to 0.02 to 0.05 µg/ml by 4 to 6 hours; in rats drug
was not detectable 180 minutes after injection.[92] The availability of tritium-labeled vinblastine[93-95] offered the possibility of more intensive study of the metabolic fate of this
compound. Within 30 minutes of intravenous injection into rats,
less than 1.5 percent of the administered radiolabeled vinblastine was found in the blood. Over the course of a 24-hour
period, only 6.63 percent of the tritium was excreted in the
urine, compared with 20-25 percent, of which only 2 percent was
unchanged drug, in the bile.[94] After intraperitoneal injection
of tritiated vinblastine, peak levels of radioactivity in the
blood were attained in 1 to 2 hours; 75 percent of this was
associated with the buffy coat layer,[95] apparently because of
an affinity of the drug for platelets.[96]

CONCLUSION

In this survey emphasis has been placed on the changes
produced in cells by exposure to the *Catharanthus* alkaloids.
Despite the variety of these changes, it is appropriate that
an attempt be made to achieve some sort of synthesis, starting
with the microtubular interactions.

The microtubules themselves probably have as their primary
role a function in motility. This would explain their presence
in cilia, in phagocytotic cells, where they may produce salta-
tory movements,[97] in cytoplasm near cell walls, in the mitotic
spindle, and in the sieve plates of plant phloem, where they
may aid in peristaltic movement of nutrients.[98] Figure 3 presents in schematic form some idea of current concepts of microtubular assembly and function. The possible sites of action
of antimitotics upon these processes are indicated; such action

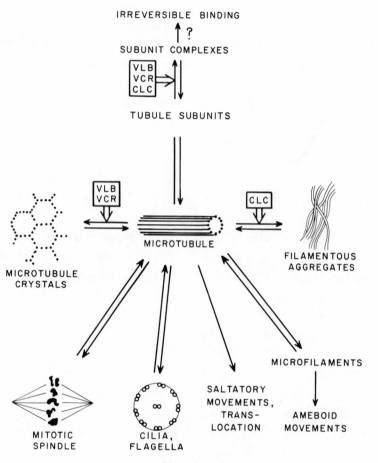

Fig. 3.

Interrelationships of the microtubule system, and the points of interaction of the *Catharanthus* alkaloids (VLB,VCR) and colchicine (CLC). The actual arrangement of the microfibrils within the microtubular structure could be spiral.[62]

could be at the level, either of the monomeric microtubule protein, or at that of protein when it is incorporated into complete tubules. The processes are envisaged as being in a state of equilibrium, an equilibrium that can be shifted to suit the requirements of cell function, as for example when spindle formation is initiated at the beginning mitosis. Binding interactions would disturb this general equilibrium, especially if there were formation of a more tightly-bound component, as some studies would indicate.[17,74] Since, however, only a small, time-dependent fraction of the total binding may be irreversible, we have an explanation for the fact that in general the cytological changes are reversible, unless exposure to the drug has been prolonged.

Cytotoxicity, on the other hand, requires that there be irreversible processes, and does not appear to be well correlated with mitotic arrest.[99] In addition, analyses of lethal action in relation to the cell cycle in HeLa and Chinese hamster cells, suggest that the S phase is the most sensitive to vinblastine and vincristine; there is also a late G1 phase sensitivity to vinblastine.[100] Such data underline the importance of effects on biosynthetic pathways in the overall cytotoxic action. Conceivably, under these circumstances the inhibitory biochemical effects could both reduce cellular proliferation, in the same way as other chemotherapeutic antimetabolites, and also reinforce the mitotic arrest due to binding of microtubule subunits by preventing their resynthesis. Such a dual action, represented in Figure 4, which would involve elements of what has been termed "complementary inhibition"[101], might constitute an effective cytotoxic mechanism for the *Catharanthus* alkaloids.

Finally, it is appropriate to briefly discuss the possible directions that future studies with these compounds might take. First, studies of the mechanism of action could be expected increasingly to focus on the molecular basis of the interaction with microtubule subunits. Knowledge of this would serve primarily as a tool with which to approach the problem of the function of these organelles in cell processes. Secondly,

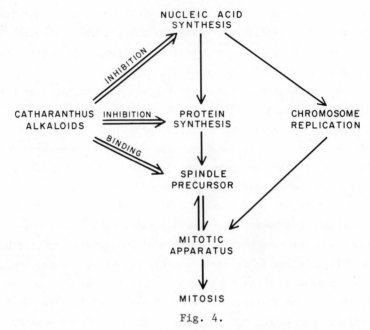

Fig. 4.

Sites of action of the *Catharanthus* alkaloids on the mitotic process.

efforts are being made to obtain new derivatives of these agents by chemical modification. Recent examples are the 4-acyl derivative of vinblastine, vinglycinate[102], and dihydrovinblastine.[103] Such developments, which are likely to continue, are of great importance to clinical cancer chemotherapy, where they may serve the twofold aim of introducing effective new agents, and of circumventing the problem of drug resistance.

ADDENDUM

Although more than two years have elapsed since this chapter was written, the advances that have been made in our understanding of the mechanisms of action of these alkaloids have only served to underline and better delineate the role of microtubule interactions. The purified microtubule protein now appears to be a heterodimer with three distinct types of active site, distributed in an as yet undefined manner between the subunits.[104,105]

Each molecule contains one site for binding of colchicine and another for *Catharanthus* alkaloids, while guanosine triphosphate interacts at two sites, one exchangeable the other nonexchanging; it is possible that one of these latter two sites is the alkaloid site. Neither the binding of vinblastine nor aggregation into microtubule crystals appear to block the colchicine site, since the crystals may still contain tritiated colchicine.[106] The biological significance of these interactions is emphasized by the finding that in Tetrahymena, adaptation to the cytotoxicity of colchicine is accompanied by reassembly of the microtubular structures initially disrupted by the drug.[107] Finally, in the area of metabolism and distribution of these drugs, concentration of vinblastine in leukocytes, extensive degradation and a primarily fecal route of excretion have been demonstrated in dogs.[108]

REFERENCES

1. J.H. Cutts, Proc. Am. Assoc. Cancer Res., 2, 289 (1958).

2. J.H. Cutts, C.T. Beer, and R.L. Noble, Cancer Res., 20, 1023 (1960).

3. I.S. Johnson, H.F. Wright, and G.H. Svoboda, J.Lab.Clin. Med., 54, 830 (1959).

4. G.H. Svoboda, J.Pharm.Sci., 47, 834 (1958).

5. G.H. Svoboda, Lloydia, 24, 173 (1961).

6. N. Neuss, I.S. Johnson, J.G. Armstrong, and C.J. Jansen, Adv. Chemother., 1, 133 (1964).

7. J.H. Cutts, Cancer Res., 21, 168 (1961).

8. G. Cardinali, G. Cardinali, A.H. Handler and M.F. Agrifoglio, Proc. Soc. Exptl. Biol. Med., 107, 891 (1961).

9. C.G. Palmer and A.K. Warren, Proc.Am.Assoc. Cancer Res., 3, 350 (1962).

10. I.S. Johnson, H.F. Wright, G.H. Svoboda and J. Vlantis, Cancer Res., 20, 1016 (1960).

11. J.H. Cutts, Canad, Cancer Conf., 4, 363 (1961).

12. J.H. Cutts, Biochem. Pharmacol. 13, 421 (1964).

13. I.S. Johnson, J.G. Armstrong, M. Gorman, and J.P. Burnett, Jr., Cancer Res., 23, 1390 (1963).

14. C.T. Beer, Canad. Cancer Conf., 4, 355 (1961).

15. W.A. Creasey and M.E. Markiw, Biochem. Pharmacol., 13, 135 (1964).

16. R. Love, Exptl. Cell. Res., 33, 216 (1964).

17. W.A. Creasey, Cancer Chemother. Rept., 52, 501 (1968).

18. E.K. Wagner and B. Roizman, Science, 162, 569 (1968).

19. W.A. Creasey and M.E. Markiw, Biochim. Biophys. Acta 87, 601 (1964).

20. W.A. Creasey and M.E. Markiw, Biochim. Biophys. Acta, 103, 635 (1965).

21. J.F. Richards, R.G.W. Jones, and C.T. Beer, Cancer Res., 26 876 (1966).

22. R.G.W. Jones, J.F. Richards, and C.T. Beer, Cancer Res., 26, 882 (1966).

23. J.L. Van Lancker, A.L. Flangas, and J. Allen, Lab. Invest., 15, 1291 (1966).

24. A. Luyckx and J.L. Van Lancker, Lab. Invest., 15, 1301 (1966).

25. P.H. Wiernik and R.M. Macleod, Proc. Soc. Exptl. Biol. Med., 119, 118 (1965).

26. R. Desjardins, D.E. Grogan, J.P. Arendell, and H. Busch, Cancer Res., 27, 159 (1967).

27. I.J. Slotnick, M. Dougherty, and D.H. James, Jr., Cancer Res. 26, 673 (1966).

28. M.T. Hakala, Fed. Proc., 28, 856 (1969).

29. M.J. Cline, Brit. J. Haematol., 14, 21 (1968).

30. W.A. Creasey, K.G. Bensch, and S.E. Malawista, Biochem. Pharmacol, in press.

31. B.M. Agustin and W.A. Creasey, Nature, 215, 965 (1967).

32. L. Chung and J.D. Gabourel, Fed. Proc. 27, 760 (1968).

33. W.A. Creasey, Biochem. Pharmacol., 18, 227 (1969).

34. P.L. McGeer and E.G. McGeer, Biochem. Pharmacol., 12, 297 (1963).

35. W.A. Creasey and M.E. Markiw, Proc. Intern. Cong. Biochem. 6th, New York (1964), Abstracts. p. 38.

36. M.J. Cline, Clin. Res., 15, 334 (1967).

37. W.A. Creasey, Fed. Proc., 27, 760 (1968).

38. P. Warnecke and S. Seeber, Z. Krebsforsch., 71, 361 (1968).

39. J. Ilan and J.H. Quastel, Biochem. J., 100, 448 (1966).

40. M.A. El-Nakeeb and J.O. Lampen, Biochem. J., 92, 59 P (1964).

41. O. Warburg, Metabolism of Tumors, Smith, New York (1931).

42. J.C. Hunter, Biochem. Pharmacol., 12, 283 (1963).

43. P. Obrecht and N.E. Fusenig, European J. Cancer, 2, 109 (1966).

44. B.J. Katchman, R.E. Zipf, and J.P.F. Murphy, Clin.Chem., 9, 511 (1963).

45. W.S. Bullough, Exptl. Cell. Res., 1, 410 (1950).

46. J.E. Amoore, J. Cell Biol., 18, 555 (1963).

47. R.H. Adamson, R.L. Dixon, M. Ben, L. Crews, S.B. Shohet, and D.P. Rall, Arch. Int. Pharmacodyn., 157, 299 (1965).

48. Q.L. Uy, T.H. Moen, R.J. Johns, and A.H. Owens, Jr., Johns
     Hopkins Med. J., 121, 349 (1967).

49. P.G. Gottschalk, P.J. Dyck, and J.M. Kiely, Neurology, 18,
     875 (1968).

50. C.J. Kensler, Cancer Res., 23, 1353 (1963).

51. D.W. Yesair, F.A. Kohner, W.I. Rogers, P. Baronowsky, and
     C.J. Kensler, Cancer Res., 26, 202 (1966).

52. G.L.A. Graff, C. Gueuning, and J. Hildebrand, Compt. Rend.
     Soc. Biol. (Paris), 161, 2645 (1967).

53. E. Mascitelli-Coriandoli and P. Lanzani, Arzneimittel-
     Forsch., 13, 1011 (1963).

54. M. Bariety and A. Gadjos, Presse Med., 73, 921 (1965).

55. J.W. Moncrief and K.S. Heller, Cancer Res., 27, 1500 (1967).

56. J. Padawer, J. Cell Biol., 29, 176 (1966).

57. E. Gillespie, R.J. Levine, and S.E. Malawista, J. Pharmacol.
     Exptl. Therap., 164, 158 (1968).

58. S.E. Malawista, M. Sato, and K.G. Bensch, Science, 160,
     770 (1968).

59. C. Rizzoli, L. Simonelli, and M. Gennari, Boll. Soc. Ital.
     Biol. Sper., 40 119 (1964).

60. P. George, L.J. Journey, and M.N. Goldstein, J. Natl.
     Cancer Inst., 35, 355 (1965).

61. A. Krishan, Fed. Proc., 27, 670 (1968).

62. K.G. Bensch and S.E. Malawista, J. Cell. Biol., 40, 95
     (1969).

63. S.S. Schochet, Jr., P.W. Lambert, and K.M. Earle, J.
     Neuropath. Exptl. Neurol., 27, 645 (1968).

64. S.E. Malawista, H. Sato, W.A. Creasey, and K.G. Bensch, Fed.
     Proc., 28, 875 (1969).

65. J.G. White, Amer. J. Pathol., 53, 447 (1968).

66. S.E. Malawista, Arthritis Rheumatism, 11, 191 (1968).

67. A.C. Sartorelli, and W.A. Creasey, Ann. Rev. Pharmacol., 9, 51 (1969).

68. E.W. Taylor, J. Cell Biol., 25, 145 (1965).

69. G.G. Borisy and E.W. Taylor, J. Cell Biol., 34, 525 (1967).

70. G.G. Borisy and E.W. Taylor, J. Cell Biol., 34, 535 (1967).

71. M.L. Shelanski and E.W. Taylor, J. Cell Biol., 34, 549 (1967).

72. R.C. Weisenberg, G.G. Borisy, and E.W. Taylor, Biochemistry, 7, 4466 (1968).

73. L. Wilson and M. Friedkin, Biochemistry 6, 3126 (1967).

74. W.A. Creasey, Pharmacologist, 9, 192 (1967).

75. W.A. Creasey and T.C. Chou, Biochem. Pharmacol., 17, 477 (1968).

76. W.A. Creasey, K.G. Bensch, and S.E. Malawista, Fed. Proc., 28, 362 (1969).

77. J.G. Georgatsos, T. Karemfyllis, and A. Symeonidis, Biochem. Pharmacol., 17, 1485 (1968).

78. R.C. Weisenberg and E.W. Taylor, Fed. Proc., 27, 299 (1968).

79. R. Marantz and M. Shelanski, J. Cell Biol., 44, 234 (1970).

80. R. Marantz, M. Ventilia and M. Shelanski, Science, 165, 498 (1969).

81. K.G. Bensch, R. Marantz, H. Wisniewski and M. Shelanski, Science, 165, 495 (1969).

82. R. Weisenberg and S. Timasheff, Biophys. J., 9, 174A (1969).

83. J.B. Olmsted, K. Carlson, K. Klebe, F. Ruddle, and J. Rosenbaum, Proc. Natl. Acad. Sci. U.S., 65, 129 (1970).

84. L. Wilson, J. Bryan, A. Ruby, and D. Mazia, Proc. Natl. Acad. Sci., U.S., 66, 807 (1970).

85. P. Harris, J. Cell Biol., 14, 475 (1962).

86. B. Afzelius, J. Biophys. Biochem. Cytol., 5, 269 (1959).

87. H.V. Rice and W.M. Laetsch, Am. J. Bot., 54, 856 (1967).

88. S.E. Malawista and K.G. Bensch, Science, <u>156</u>, 521 (1967).

89. A.C. Taylor, J. Cell Biol., <u>28</u>, 155 (1966).

90. N.K. Gonatas and E. Robbins, Protoplasma, <u>59</u>, 377 (1965).

91. R.L. Noble, Canad, Cancer Conf., <u>4</u>, 333 (1961).

92. G.J. Dixon, E.A. Dulmadge, L.T. Mulligan and L.B. Mellett, Cancer Res., <u>29</u>, 1810 (1969).

93. P.E. McMahon, Experientia, <u>19</u>, 434 (1963).

94. C.T. Beer, M.L. Wilson, and J. Bell, Canad, J. Physiol. Pharmacol., <u>42</u>, 368 (1964).

95. H.F. Greenius, R.W. McIntyre, and C.T. Beer, J. Med. Chem. <u>11</u>, 254 (1968).

96. H.F. Hebden, J.R. Hadfield, and C.T. Beer, Cancer Res., <u>30</u>, 1417 (1970).

97. J.J. Freed, A.N. Bhisey, and M.M. Lebowitz, J. Cell. Biol. <u>39</u>, 46 A (1968).

98. R. Thaine, Nature, <u>222</u>, 873 (1969).

99. E. Frei, III, J. Whang, R.B. Scoggins, E.J. Van Scott, D.P. Rall, and M. Ben, Cancer Res., <u>24</u>, 1918 (1964).

100. H. Madoc-Jones and F. Mauro, J. Cell Physiol. <u>72</u>, 185 (1968).

101. A.C. Sartorelli, Nature, <u>203</u>, 877 (1964).

102. I.S. Johnson, W.W. Hargrove, P.N. Harris, H.F. Wright, and G.B. Boder, Cancer Res., <u>26</u>, 2431 (1966).

103. R.L. Noble, C.T. Beer, and R.W. McIntyre, Cancer, <u>20</u>, 885 (1967).

104. J. Bryan, Biochemistry, <u>11</u>, 2611 (1972).

105. R.C. Weisenberg, G.G. Borisy, and E.W. Taylor. Biochemistry, <u>7</u>, 4466 (1968).

106. A. Krishan and D. Hsu, J. Cell Biol., <u>48</u>, 407 (1971).

107. F. Wunderlich and V. Speth, Protoplasma, <u>70</u>, 139 (1970).

108. W.A. Creasey and J.C. Marsh, Proc. Amer. Assoc. Cancer Res., <u>14</u>, 57 (1973).

CHAPTER VIII

CLINICAL ASPECTS OF THE DIMERIC CATHARANTHUS ALKALOIDS

R.C. DeConti and W.A. Creasey

Departments of Internal Medicine and Pharmacology
Yale University, School of Medicine
New Haven, Connecticut 06510

INTRODUCTION

In the comparatively short time that has elapsed since
their introduction into the clinic, the dimeric alkaloids from
Catharanthus roseus (*Vinca rosea*) have earned themselves a
place among the most valuable agents used in cancer chemotherapy. Not only have they proven useful as single agents
in the palliative treatment of several advanced neoplasms, but
more recently they have been utilized in various schedules of
combination chemotherapy that may achieve a higher order of
therapeutic success. The sheer bulk of the clinical literature
regarding the *Catharanthus* alkaloids forbids anything approach-
ing a comprehensive review of the area. Therefore, it is our
intention in this chapter to make selective use of data from
our own and other centers so as to give a concise overall picture of the clinical status of these drugs. We shall outline
the therapeutic applications, the type of clinical benefits
achieved and the side effects encountered; individual case
histories will be included for illustration.

237

FORMULATIONS

It is appropriate to begin by listing those alkaloids that
have received clinical trials (Table I); of these only vinblastine and vincristine are of established value. Oral
preparations of vinblastine have been introduced to avoid the
necessity for intravenous injection, but the uncoated tablets
at least have been subject to problems of irregular absorption and gastrointestinal toxicity that have forced their withdrawal. Clinical trials with the enteric coated preparations
are underway and may avoid some of these difficulties.[2] Vinleurosine, as the methiodide[3] and as the sulfate,[4,5] has been
administered to more than 100 patients. Leukopenia has been
less pronounced than with vinblastine, but a shock-like syn-
drome has accompanied rapid intravenous injection, and the
therapeutic benefits have been notably inferior to those of the
other alkaloids. The fourth naturally-occurring alkaloid,
vinrosidine, despite promising preclinical results,[1] was found
to have unacceptable human toxicity without therapeutic benefit.
[5] The remaining alkaloid, vinglycinate, is the result of a
program to obtain new agents by chemical modification of existing compounds. In clinical trial,[6] it has shown promise in
the treatment of Hodgkin's disease, lymphosarcoma and broncho-
genic carcinoma. Its spectrum of activity and limiting toxicity resemble those of vinblastine, but a ten-fold higher dose
level is required. Of great interest is the reported lack of
cross-resistance to vinblastine and vincristine. Clinical
trials of this drug have been halted temporarily because of
problems with the stability of current preparations.[7]

CLINICAL PHARMACOLOGY

Only two aspects of the human biochemistry and pharmacology
of the *Catharanthus* alkaloids have been explored. These include studies of the plasma lifetimes of injected drugs, and

TABLE I

Catharanthus Alkaloids and Derivatives that have Received Clinical Trial

Generic Name	Alternate Name	Form	Brand Name	NSC Number	Preparations Available	Clinical Status
Vinblastine	Vincaleukoblastine	Sulfate	Velban[R] Velbe[R]	49842	Ampoules – 10 mg	Generally available
					Tablets, uncoated– 5, 10, 20 mg	Discontinued
					Tablets, enteric coated – 5 mg	Investigational use
Vincristine	Leurocristine	Sulfate	Oncovin	67574	Ampoules – 1, 5 mg	Generally available
Vinleurosine	Leurosine	Sulfate Methiodide		90636	Ampoules – 100 mg	Discontinued Discontinued
Vinrosidine	Leurosidine	Sulfate				Discontinued
Vinglycinate	Desacetylvin- blastine-4-(N,N- dimethylglycinate)	Sulfate				Investigational use

experiments designed to develop biochemical tests to predict or
monitor therapeutic response. The development of assay systems
involving KB[8] or L cells,[9] has enabled determinations to be
made of the blood levels of vinblastine or vincristine, although
of course it has not provided information on metabolic transformations. With vinblastine, injection of 0.2 mg/kg produced
serum levels of less than 0.05 µg/ml,[8] while in children with
acute leukemia treated with vincristine (0.1 mg/kg), partial
inhibition of cell proliferation, corresponding to drug levels
of 0.005 to 0.05 µg/ml, persisted for as long as 4 hours in
50% of the subjects tested.[10]

Reports of inhibition of RNA synthesis in human neoplastic
cells by treatment with vinblastine[11,12] and vincristine[13] have
formed a basis for tests carried out *in vitro* with leukemic
cells. Inhibition of uridine incorporation into RNA occurred
with vincristine and other antineoplastic agents, the extent
of this effect being correlated with cytotoxic action *in vivo*,
although the system did not appear to predict true hematologic
remission.[14,15] On the other hand, in a study in which changes
in trypan blue staining behavior were used as an index of sensitivity of leukemic cells to drugs *in vitro*, there was a close
correlation between this index and the incidence of complete
remissions in children with acute leukemia who received vincristine, daunorubicin and prednisolone in combination.[16] In
patients with acute granulocytic leukemia treated with vincristine, several enzymic activities and the incorporation of
precursors into DNA were studied in isolated leukocytes for a
48 hour period after drug; the general response seemed to be
an early elevation followed by a depression at 24 and 48 hours.
[17] The activities of thymidylate synthetase, dihydrofolic
reductase and incorporation of deoxynucleosides into DNA were
closely correlated together, and the levels of thymidine and
uridine kinases also changed as a unit, but independently of
the other parameters.

It is evident that much work needs to be done in the area of the clinical pharmacology of the *Catharanthus* alkaloids. The availability of tritium-labeled vinblastine offers an opportunity to explore in detail the distribution and metabolic fate of this compound.

ADMINISTRATION AND TOXICITY

The method of administration and the dosage levels used for these alkaloids, like many other antineoplastic agents, are closely tied to their clinical toxicity. Careful titration of the dose to achieve minimal toxicity consistent with therapeutic benefit is necessary for successful clinical use.

Vinblastine, the first alkaloid to be introduced into the clinic, exhibits the major side-effects listed in Table II. It is noteworthy that vinblastine, although causing leukopenia, does not produce a significant incidence of thrombocytopenia, indeed thrombocytosis has been reported.[21] Phlebitis and cellulitis at the injection site result from the local irritant action of high drug concentrations; they are minimized by ensuring that the injection needle is well seated in the vein and by washing the needle with venous blood before withdrawal. In addition to the listed effects, a variety of other toxic symptoms have been noted in a relatively few patients. These include pain in the primary tumor,[20] hemorrhagic enterocolitis,[20] salivary-gland pain,[22] and a skin reaction in areas previously irradiated resembling that seen with actinomycin D.[23]

In contrast to the situation with vinblastine where hematologic toxicity is dose-limiting and neurologic toxicity is generally mild or appears only after prolonged therapy, vincristine has major neurologic effects as its limiting feature and is only a mild bone marrow depressant.

A minor substitution on a complicated chemical structure has not only changed the spectrum of clinical usefulness but

TABLE II

Major Toxic Manifestations of Parenteral Vinblastine Therapy

Side Effect	Number of Patients	%
Leukopenia (WBC < 2,000/mm^3)	160	31.6
Nausea, vomiting and anorexia	86	17.0
Neurotoxicity (Loss of reflexes, paresthesias, peripheral neuritis)	42	8.3
Epilation	33	6.5
Stomatitis	17	3.4
Diarrhea	12	2.4
Constipation	11	2.2
Malaise, lethargy and depression	11	2.2
Phlebitis at injection site	11	2.2
Ileus	8	1.6
Rectal bleeding	4	0.8
Hematuria	2	0.4
Thrombocytopenia (< 100,000/mm^3)	2	0.4

Data derived from published figures on 506 patients.[18-20]

created a new feature of toxicity at doses approximately one-tenth those of vinblastine. While individual doses of 2 mg/M^2 (approximately 0.078 mg/kg) are reasonably well tolerated by children, such dosages in adults may produce acute symptoms of central nervous system toxicity with alterations in consciousness, convulsions and psychosis. In addition, single such doses may lead to profound peripheral neuropathy. Dosages of less than 0.05 mg/kg have been recommended,[24] and some centers limit total individual dose to a maximum of 2 mg per injection. Like vinblastine, vincristine is administered intravenously at 1-2 week intervals and care must be taken to avoid subcutaneous

infiltration. To minimize toxicity, the dosage in adults may
be started at 0.02 mg/kg and gradually increased to 0.05 mg/kg
per week until response or evidence of toxicity occurs. This
gradual increase allows an opportunity to reduce the dosage
before severe and/or irreversible neurologic damage occurs. The
first symptom of toxicity is usually tingling of the distal
extremities (most commonly the finger tips) followed by numb-
ness. Painful paresthesias, clumsiness and uncoordinated movements, wide-based gait, diffuse muscle weakness and specific
motor nerve palsies develop. Diplopia may signal extraocular
muscle weakness. Hoarseness may indicate laryngeal muscle
impairment. The rapidity of onset and severity of these symptoms depend on the individual as well as the total drug dose.
Modification of the dosage schedule can usually prevent the
more severe manifestations of toxicity. Table III lists the
neurologic complications and toxicity in the initial 62 patients
treated with vincristine at Yale.[25] While the sensory defects
of peripheral neuropathy occurred commonly, motor weakness and
palsy were relatively infrequent. Patients in this study
received vincristine at doses of 0.02-0.03 mg/kg.

Neurophysiological studies by Tobin and Sandler have demonstrated that depression and loss of the Achilles deep tendon
reflex is the earliest and most consistent sign of vincristine-
induced neuropathy; this was found in all of 50 leukemic patients
who had received a total dose of 2-4 mg/M^2. Their studies suggested a selective effect of the drug on the muscle spindle, its
gamma innervation or its annulospiral endings.[26,27] Nerve biopsy in patients with advanced neuropathy has demonstrated axonal
degeneration of nerve fibers.[28]

Severe peripheral neuropathy has been observed on a number of occasions after relatively small doses of vincristine
in patients with preexistent hepatic insufficiency.[28,29,30]
This has been postulated to be the consequence of impaired hepatic detoxification and/or biliary excretion.[1]

TABLE III

Untoward Effects of Vincristine Therapy

	Number	%
Total Patients	62	100.0
None	18	30.9
Paresthesias: Upper Extremity	30	51.6
Lower Extremity	6	10.3
Decreased or absent deep tendon reflexes	18	30.9
Constipation	10	16.1
Hoarseness	8	12.8
Generalized weakness	5	8.0
Slapping gait	4	6.4
Hair loss or Alopecia	5	8.0
Abdominal pain	3	4.8
Leg cramps	2	3.2
Mental depression	2	3.2
Phlebitis	2	3.2
Cranial Nerve Palsy	2	3.2
Jaw Pain	1	1.6
Perforated Colon	1	1.6

Hyponatremia associated with vincristine neurotoxicity has been reported at least twice with inappropriate antidiuretic hormone secretion.[31,32]

Alopecia is a more frequent consequence of vincristine than vinblastine therapy and depends upon the dosage and duration of therapy. Some hair loss may occur in the majority of treated patients. This occurred in only 8 % of our initial patients. Recovery occurs despite continued therapy though 3-4 months are required for noticeable improvement. The use of a headband

inflated to arterial pressure for 15 minutes may be a useful
means of reducing this complication. Several studies attempting to improve the therapeutic index of the *Catharanthus* alkaloids
by intraarterial infusion of these drugs to localized tumors
have not been encouraging. Severe skin reactions may have been
the result of susceptibility to extravasation as well as high
concentrations of drug.[33,34] Therapeutic efficacy remains in
doubt.[35]

The occurrence of meningeal leukemia despite successful
vincristine-prednisone induction therapy in acute childhood
leukemia has suggested poor penetration of the blood brain
barrier by the drug and need for intrathecal therapy. However,
animal toxicity data, the known central nervous system effects
of the drug, and one unfortunate clinical experience[36] clearly
contraindicates the intrathecal use of vincristine.

Other Biologic Effects

In contrast with the often detailed information which is
available regarding a number of biologic properties of other
classes of antineoplastic agents, knowledge of these effects
of the *Catharanthus* alkaloids in man is fragmentary.

Many antineoplastic agents have been found to be immun-
osuppressive in man and in a variety of test systems.[37] Animal
systems have been used to demonstrate this ability for the
Catharanthus alkaloids,[38,39] and two studies have been reported in
man. Vinblastine has been shown to inhibit the lymphocyte
response to phytohemagglutinin *in vitro*,[40] and combination
therapy with vincristine and prednisone resulted in inhibition
of the primary immune response to tularemia, Vi and pneumococ-
cal polysaccharide antigens.[41] Inasmuch as prednisone has
never really been shown to have immunosuppressant activity in
man, the effects are probably ascribable to vincristine which
was administered at doses of 2 mg/M^2/week.

Suppression of ovarian function[42] and alterations in spermatogenesis[43] are best documented in patients receiving alkylating agents. Amenorrhea has been noted in one patient receiving vinblastine therapy alone,[44] and changes in spermatogenesis have been observed in rats.[45]

Marked enbryocidal and teratogenic effects have been produced in golden hamsters[46] by vinblastine and vincristine, and camparable effects have been achieved in the rat[47] with vinblastine. To date fetal abnormalities have not been described in man though several normal offspring of vinblastine-treated mothers have been reported.[48-50] Extreme caution should be exercised in interpreting these data as suggesting that *Catharanthus* alkaloids are safe in pregnancy however, in view of the effects of other antineoplastic agents in man.[51]

CLINICAL APPLICATION
Lymphoma

The major use of vinblastine in cancer chemotherapy is in the management of patients with Hodgkin's disease. In patients with Stage IIIB or IV disease who are no longer candidates for high dose, extended field radiotherapy with curative intent, vinblastine closely rivals nitrogen mustard (Mechlorethamine Hydrochloride) as the initial choice in single agent chemotherapy. Responses are comparable with both agents producing useful benefits in 70-80% of patients. Complete remission defined as complete disappearance of all objective evidence of disease occurs in about one-third of all responders to both agents; partial remission, that is some reduction in objective manifestations of disease with subjective improvement, occurs in the remainder. One randomized study[52] of these 2 agents as initial therapy demonstrated objective response in 80% of the patients treated with nitrogen mustard, compared to 73% objective responses with vinblastine. The median duration of re-

sponse was 2.5 months in this study during which vinblastine
was given for 6 weeks.   The use of nitrogen mustard followed
by maintenance therapy with chlorambucil has been shown to
prolong the mean remission duration from 11 to 33 weeks.[53]
Similarly, the continued administration of vinblastine at one-
to three-week intervals achieved objective responses with a
7 month median duration (Figure 1). Subjective improvement
was sustained for 13 months. One-third of those responding
to vinblastine were maintained in remission for more than one
year; occasional two-year responses occurred[54] and at least
one five year response has been reported.[55] Studies of the
efficacy of vinblastine compared with cyclophosphamide demonstrated the superiority of vinblastine over this alkylating
agent.[56,57] One study has compared the response rate of alkylating agent resistant patients with vinblastine and procarbazine.
100% of the patients receiving procarbazine responded as compared to only 68% with vinblastine. However, the mean duration

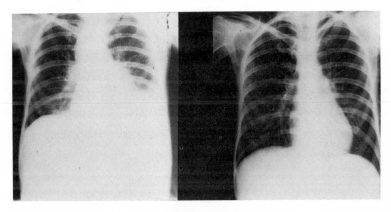

Fig. 1

Serial Chest X-rays demonstrating resolution of pleural
effusion and mediastinal widening 4 weeks after vinblastine
therapy for Hodgkin's disease.

of vinblastine response was 10.5 months as compared with 6
months for procarbazine.[58] This lack of cross resistance makes
this agent of inestimable value in prolonging the total useful
remission time which may be achieved with drug therapy.

Vinblastine therapy is easily monitored in practice by
checking the patient's response and the white blood cell count
prior to each dose. Thrombocytopenia is a rare manifestation
of hematologic depression and the white cell count alone is
usually an accurate guide to therapy. Drug treatment is delayed, or the dosage is reduced, if moderate leukopenia (WBC <
3000) occurs. Therapy is usually initiated at doses of 0.1 mg/
kg and may be raised by 0.05 mg/kg increments weekly to a maximum of 0.3 mg/kg if response is not achieved. Response is
usually evident at 3 weeks, though maximal benefits may require
6 weeks of therapy. Once remission is achieved, the intervals
between injections may gradually be lengthened to 10-14 days
without harm.

Vinblastine therapy is conveniently carried out in an outpatient clinic. This avoids the necessity for hospital admis-
sion, the acute nausea and vomiting, and a potential period of
prolonged granulocytopenia, which are frequent concomittants of
intravenous alkylating agent usage.

Whereas the improvements induced by vinblastine in Hodgkin's disease clearly rival those of alkylating agents, the
benefits of vincristine are of lesser magnitude. While 40-60%
of patients respond, the duration of remission tends to be
shorter and with continued therapy, development of neurologic
side effects becomes a virtual certainty. Nevertheless, vincristine remains a very useful agent, particularly in situa-
tions where leukopenia and thrombocytopenia as a result of
X-ray therapy or prior myelosuppressive agents, makes effective
therapy with other drugs unlikely. In contrast with alkylating
agents which are generally cross-resistant, patients resistant
to vinblastine may be expected to respond to vincristine.

Table IV describes the initial Yale experience[?]
In this study, patients resistant to alkylatin
vinblastine received at least 3 weekly doses (
of vincristine. Though 8 of 15 patients with
responded -- one for as long as 2 years -- th
of response was only 6 weeks.

In monitoring vincristine therapy, one obtains blood counts
at weekly intervals to check both for leukopenia and the development of anemia which may occur after continued therapy as a
result of suppression of erythropoiesis by vincristine.[59] More
importantly, one carefully questions the patient to elicit symptoms of developing neuropathy. Tingling of the fingers and toes
is the most frequent early symptom, followed by paresthesias,
aching pains in the limbs, impaired fine movements of the fingers, and in more advanced cases, gross muscle weakness. Reduction in dosage or increasing the interval between injections
will prevent progression of neurologic deficits and in most

TABLE IV

Beneficial Effects of Vincristine

Disease	No. Responders/ No. Patients	Duration of Response Range	Median (weeks)	Mean
Hodgkin's Disease	8/15	4-102	6	23
Lymphosarcoma	6/13	4-32	9	18
Reticulum Cell Sarcoma	6/8	5-13	7	9
Acute Leukemia	5/7	4-13	4	7
Multiple Myeloma	1/3	16	16	16
Solid Tumors	0/16			
Total	26/62			

patients symptoms slowly resolve. In the physical examination,
one pays particular attention to the activity of the deep tendon reflexes which will be first depressed and then abolished
by drug therapy. In practice, one does not resort to electro-
myographic studies of muscle function to demonstrate the development of neuropathy.

The toxic manifestations of vincristine therapy on the
autonomic nervous system include constipation and even paraly-
tic ileus. Patients should be advised to liberalize their
fluid and bulk intake and may be treated prophylactically with
hydrophilic agents to promote stool bulk. As with vinblastine,
vincristine may be administered conveniently in an outpatient
facility.

Vinblastine has been relatively ineffective in lymphosarcoma and reticulum cell sarcoma and though occasionally
beneficial its use is reserved until other potential agents have
been exhausted. Vincristine on the other hand, is quite effective therapy in these conditions. Lymphosarcoma may be a quite
indolent process, well controlled by oral alkylating agent
therapy, and in this situation the toxicity of vincristine
negates its use. However, in the undifferentiated or large cell
lymphosarcoma with fulminant spread behaving more like reticulum cell sarcoma, the drug may be almost equal to the alkylating agents in efficacy. As demonstrated in our initial experience with vincristine (Table IV), there is a high rate of response among the lymphomas, but the remissions are of very short
duration. Complete remissions are rare and most patients relapse in 6-8 weeks. Previous treatment with other forms of
chemotherapy does lower the response rate. In a comparative
study of cyclophosphamide and vincristine, Carbone and Spur
demonstrated a response rate of 71% for cyclophosphamide and
46% for vincristine in patients with lymphosarcoma. Previously
treated patients achieved a response 57 and 17% of the time,

respectively. Despite these differences, the survival of patients undergoing vincristine therapy slightly exceeded that
for patients who received cyclophosphamide as initial treatment.
In patients with reticulum cell sarcoma, response rates of 52
and 37%, respectively, were noted.[57] These data indicate that
vincristine has effective antineoplastic activity but that its
duration is brief. These considerations have led to attempts
at combination therapy to improve its therapeutic index.

Before discussing combination chemotherapy, we would like
to point out that although vincristine is also effective in
Burkitt's lymphoma, like the other agents employed against this
sensitive tumor, results depend on the size of the tumor mass
and the extent of spread.[60] More extensive trials of therapy
in this disease will be necessary to assign priorities to a
particular agent or scheme of therapy.

Combination Therapy in Lymphoma

Combination therapy in Hodgkin's disease was first initiated using vinblastine and chlorambucil[61,62] which appeared to
produce responses superior to those obtained previously with
either agent alone   The number of patients studied was small
and toxicity at least was modest. More recently, DeVita, *et
al.*[63] have demonstrated in a series of 43 patients an 81% incidence of complete remissions with durations markedly in excess of those previously reported. This combination consists
of nitrogen mustard, vincristine, prednisone and procarbazine.
The improved survival figures for this group of patients suggests a new level of accomplishment in cancer chemotherapy,
and adds to the accumulated evidence[64] that drug therapy has
prolonged survival in this disease. Our own group has been
evaluating a 5 drug-combination chemotherapy program[65,66] which
utilizes the drugs and schedule depicted in Figure 2. Vincristine with its mild hematologic toxicity is used during the

HODGKIN'S DISEASE COMBINATION THERAPY

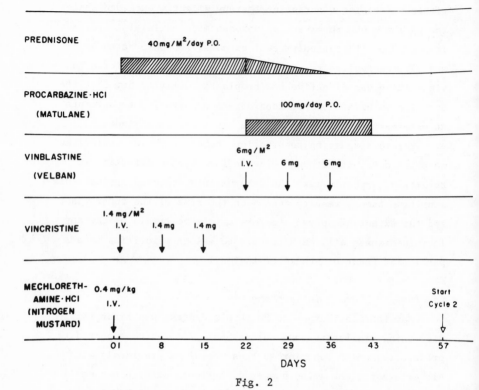

Fig. 2

The Combination Chemotherapy Drug Sequence used for
Hodgkin's Disease by the Oncology and Cancer Chemotherapy
Section at Yale.

period of marrow suppression induced by nitrogen mustard. After
hematologic recovery, vinblastine and procarbazine are added.
Thus the combination includes 5 agents, all of which have individual activity against this disease, used in a sequence designed to achieve tolerable toxicities with the hope of increasing the therapeutic index. At least three cycles of this
combination therapy are given. To date, 18 patients have been
treated; all responded and 79% achieved complete remission.

With 16 patients still in remission, we have an average mean remission duration of 9+ months. Although the data on this program are still preliminary, they are consistent with the finding by DeVita and others (see Table V) that combination chemotherapy is increasing both induction and duration of re- missions as a result of augmented neoplastic cell kill.

TABLE V

Vinca Alkaloids In Combination Therapy Programs of Lymphoma

Agents	Disease	% Complete Remissions	Duration of Response (months)	Ref.
Vinblastine Chlorambucil	Hodgkin's Disease	63	4-12	62
Vincristine Cyclophosphamide Methotrexate Prednisone X-ray	Hodgkin's Disease	71	> 32	67
Nitrogen Mustard Vincristine Prednisone Vinblastine Procarbazine	Hodgkin's Disease	79	9+	66
Vincristine Nitrogen Mustard Procarbazine Prednisone	Hodgkin's Disease Lymphosarcoma Reticulum Cell Sarcoma	81 47 37.5	> 20 11.7+ 32+	63 68
Vincristine Cyclophosphamide Prednisone	Lymphosarcoma Reticulum Cell Sarcoma	35.2 30.8	? ?	69
Cyclophosphamide Vincristine Methotrexate- Leucovorin Cytosine Arabinoside	Reticulum Cell Sarcoma	57	?	70

Programs of combination chemotherapy for other lymphomas
also appear to be producing superior results. Vincristine is
an effective agent in lymphosarcoma and reticulum cell sarcoma
and because of its minimal hematologic toxicity has been included
in many of them. Using these combinations in the most aggres-
sive histologic type of lymphoma -- reticulum cell sarcoma --
complete remissions have been achieved in from 30-57% of patients (See Table V). While therapy with single agents produces
rates of response within this range, these are almost exclusively
only partial remissions. Carbone and Spurr, for example, achieved
responses to cyclophosphamide in 52% and to vincristine in 37%
of untreated patients but only 13% and 15%, respectively, were
judged to be complete responses. These contrasts in an aggres-
sive disease for which the median survival is from 6-9 months,
suggests that the benefits of such approaches will be considerable, and that we are only beginning to exploit the agents already available in order to achieve gains in both response
rates and in overall survival.

Acute Leukemia

Vincristine was found in early clinical studies to be markedly effective in inducing remission in acute lymphocytic lau-
kemia of childhood. Responses ranging between 40% and 57% have
been observed with vincristine alone. Vinblastine was quite
ineffective and this is perhaps the most clear-cut example of
differential tumor responses to these two alkaloids. Doses of
2 mg/M^2 have been thought to be the most efficacious and have
been extended to many combinations which have subsequently been
tried. In no other disease have the benefits of combination
therapy been more clearly demonstrated in terms of both remission induction and overall-survival, which have steadily improved during the last 10 years.[71] Two year survival figures
have risen from 10 to 75% in this decade and vastly improved

5 year survival may be projected. Table VI lists a number of
well-known combinations of drugs that have included vincristine
and indicates their value.

The ability of children to tolerate proportionately larger
doses than adults may contribute to the effectiveness of vincristine in this disorder. A dosage of 2 mg/M^2 is approximate-

TABLE VI

The Effect of Drug Combination Programs on
Remission Induction in Acute Lymphocytic Leukemia of Childhood

Agent		Schedule	% Bone Marrow Remissions	Ref.
Vincristine	2	mg/M^2/week	57	72
Vincristine	2	mg/M^2/week	72	73
Prednisone	2	mg/kg/day		
Vincristine	2	mg/M^2/week	97	74
Prednisone	100	mg/M^2/day		
Daunorubicin	60	mg/M^2/week		
Vincristine	1.5	mg/M^2/week	100	75
Prednisone	100	mg/M^2/day		
Daunorubicin	20	mg/M^2/day x 2 weeks		
Vincristine	2	mg/M^2/day 1 only	88	76
Prednisone	40	mg/M^2/day x 8		
6-Mercaptopurine	60	mg/M^2/day x 8		
Methotrexate (VAMP)	20	mg/M^2/days 1 and 4		
Vincristine	2	mg/M^2/day 1 only	94	72
Prednisone	1	gm/M^2/day x 5		
6-Mercaptopurine	500	mg/M^2/day x 5		
Methotrexate (POMP)	7.5	mg/M^2/day x 5		

ly 0.078 mg/kg, which in adults may produce acute neurologic
side effects with some frequency. Children rarely develop acute
symptoms of toxicity, but nevertheless the virtual certainty
of more progressive neurologic impairment with continued drug
administration and the poor results achieved with this compound
as maintenance therapy has tended to limit its usefulness to
induction of remission. Four to six consecutive weekly doses
of vincristine may be necessary before its effect can be established. The median response time is about 28 days. Most programs for combination therapy of acute leukemia have combined
vincristine with prednisone. Patients usually respond repeat-
edly to courses of such therapy for induction of remission[77],
[78,79] and the combination has also been used in programs of
consolidation and reinduction during the period of remission.

New   Vincristine has been somewhat less successful in acute
myelocytic leukemia in children (37% complete remission) and
is quite ineffective when used alone in adult acute myelocytic
leukemia. In selected series of multiple drug combinations,
complete remissions have been observed in 22 to 45% of patients
treated.[72] The blast crisis phase of chronic myelocytic leukemia associated with chromosomal aneuploidy has responded to
vincristine-prednisone combination therapy,[80] while chronic
myelocytic leukemia itself has been thought to be an entity
that is relatively unaffected by vincristine.[81]

New   No doubt many other studies utilizing vincristine in combination therapy of leukemia will be carried out in the future,
encouraged by the efficacy of the compound when used alone and
in multiple combinations to date. Furthermore, the minimal
marrow suppressant effects of the drug have enabled it to be
combined with agents that predominantly depress marrow function
in what appears to be a successful effort to maximize leukemia
cell kill while at the same time diversifying the toxic effects.

Solid Tumors in Children

In contrast to solid tumors in adults which are predominantly adenocarcinomas and epidermoid carcinomas, solid tumors in children are almost exclusively of embryonal or mesenchymal origin. These neoplasms tend to be responsive to antineoplastic drug treatment and in the *Catharanthus* alkaloid group, to vincristine. Beneficial objective and subjective responses have been noted in children with Wilms' tumor, neuroblastoma, rhabdomyosarcoma other soft tissue sarcomas, retinoblastoma, malignant teratoma, Ewing's sarcoma, central nervous system tumors and Histiocytosis X.[82,83]

Actinomycin D is the primary antineoplastic drug choice for Wilms' tumor; this antibiotic produces remissions frequently and when used in a sequential program of drugs, surgery and X-ray therapy has markedly prolonged the disease-free interval before recurrence and may even be responsible for some cures.[84] Vincristine is not cross-resistant with actinomycin D, produces regressions in metastatic disease in 60-75% of patient and when used in conjunction with radiation therapy may result in some prolonged survivals and cures.[85] Vincristine is generally given at doses of 2 mg/M^2 weekly for 8-12 doses and response is initially evident in 2-6 weeks.

Cyclophosphamide induces tumor regressions in about 50% of patients with neuroblastoma[86-89] and vincristine alone has produced responses in 23 to 55% of patients.[90,83] The use of these 2 agents in combination has suggested augmented effects. Table VII describes the results of 4 such combination studies with responses ranging from 25-100% and an overall response rate of 64%. Unfortunately, in addition to attempting to explain differences in dosage schedules of drug, the biology of neuroblastoma itself makes any two studies difficult to compare. Responses to drug are clearly related to young age as well as to extent of disease. Despite these difficulties, the number

TABLE VII

Vincristine-Cyclophosphamide Therapy in Neuroblastoma

Study	Responses	Comment
James, 1965[91]	9/9	A young group
Pratt, 1968[92]	11/14	No ages given
Evans, 1969[93]	9/28	Best in children < 1 year
ALGB, 1969[94]	27/47	Children < 2 longer survivals

of long-term survivors and potential cures with this combination seems to exceed that which is expected from single-agent therapy alone.

Similarly, studies with rhabdomyosarcoma have demonstrated a high rate of response (54%) to vincristine alone[83] but only of short duration (median of one month), whereas preliminary combination drug programs using vincristine together with actinomycin and cyclophosphamide suggest longer remission durations.[92]

Solid Tumors in Adults

By and large the responses of non lymphomatous epidermoid tumors or adenocarcinomas in adults have been limited. In part, this may be related to the limited doses tolerated in adults, particularly in relating to vincristine. Isolated reports of responses have been described in many tumors though response rates overall are very low.[95,96] The most striking exception to this generalization has been the response of choriocarcinoma and other related gestational tumors to vinblastine and vincristine. Patients resistant to the antineoplastic effects of methotrexate and actinomycin D have responded with complete remissions and presumed cures of this transplanted neoplasm.[97]

Useful responses to vinblastine have been reported with some frequency in epidemoid carcinomas of the head and neck,[98] some testicular tumors, carcinoma of the ovary and the breast.[99]

Vinblastine has been used with favorable effects in combination with chlorambucil in the latter two diseases.[100] Vincristine has produced benefits in patients with central nervous system neoplasms,[101] a variety of soft tissue sarcomas as well as ovarian and breast carcinoma.[102] In general, soft tissue sarcomas in adults do not seem to respond as well as childhood tumors. Both vinblastine and vincristine have produced transient tumor responses in carcinomas of the lung though response rates are low[98,103,104] for other than oat cell carcinomas.[30]

Because of its relatively slight hematologic suppressant properties, vincristine has been combined with antibiotics, alkylating agents and antimetabolites in many varieties of combination chemotherapy programs in an effort to improve the quality and duration of response in many tumors unresponsive or only minimally responsive to single agent chemotherapy (See Table VIII). Analysis of the value of such therapy is rendered difficult by inclusion of multiple tumor types in a single study as well as by the differences in the dosages and schedules of the individual agents used. However, the preliminary estimates in some of these studies are encouraging and more will need to be done. The high number of responders to a combination program of vincristine and bischlorethylnitrosourea in patients with malignant melanoma is suggestive of synergistic antineoplastic effects.[113] A complicated 5 day program including vincristine in breast cancer has produced a very striking complete remission rate[53-60] despite inclusion of patients with poor prognostic indicators.[114] Studies like these will undoubtedly extend the range of neoplasms affected by the *Catharanthus* alkaloids in the future.

TABLE VIII

Vincristine in Solid Tumor Combination Chemotherapy

Drug Combination Vincristine +	Disease	Responses	Ref.
Actinomycin	Rhabdomyosarcoma	6/6	105
Actinomycin+ Cyclophosphamide	Rhabdomyosarcoma	7/7	92
Actinomycin	Multiple Tumors	21/31	106
Mitomycin C + Melphalan	Multiple Tumors	14/53	107
Mitomycin C + 5 Fluorouracil + Triethylene thiophosphoramide	Multiple Tumors	4/37	108
Methotrexate + 5 Fluorouracil + Cyclophosphamide	Multiple Tumors	9/24	109
Methotrexate	Multiple Tumors	31/58	110
Methotrexate + Melphalan	Testicular Tumors	2/8	111
Methotrexate + 5 Fluorouracil + Cyclophosphamide	Testicular Tumors	7/17	112
Bischlorethyl-nitrosourea	Malignant Melanoma	8/18	113
Cyclophosphamide + Prednisone + 5 Fluorouracil + Methotrexate	Breast Cancer	53/60	114

ILLUSTRATIVE CASE MATERIAL

Case 1

This 21 year old white auto mechanic presented to the Yale-New Haven Hospital in December 1959 with a 2 month history of cough and fever. Cervical adenopathy and splenomegaly were noted on physical examination and mediastinal widening was found on chest X-ray. Lymph node biopsy revealed Hodgkins disease which was considered Stage III-B. A 6 day course of 6-azauridine produced only a short response and in May 1960, he was given mechlorethamine HCl with improvement that lasted only a few months. Treatment with prednisone was begun in June 1961 and led to some regression of adenopathy, but because of decreasing responsiveness, he was begun on weekly therapy with vinblastine in dosages ranging from 0.15-0.25 mg/kg. On this therapy, his constitutional symptoms gradually abated and within a month his supraclavicular adenopathy had disappeared (Figure 3). This therapy was continued almost weekly until March 1962 (6 months) without evidence of relapse. The decision to discontinue therapy was prompted by the development of back pain felt to be a neuritis. During this period his white blood cell count had transiently fallen to 2500-3500 mm^3 and a few weekly doses had been withheld. Without therapy he improved, the neuritis cleared and he was completely functional until December 1962 when he again developed adenopathy. He responded once more to vinblastine (0.2 mg/kg) given at 3 week intervals until April when adenopathy again was noticed. Decreasing the interval between injections of drug to every 2 weeks controlled the process until February 1964, without recurrence of neuritis. Subsequently he benefited from treatment with chlorambucil, local irradiation therapy and cyclophosphamide. In November 1965 he was treated with vincristine without significant response and he expired in February 1966.

Comment:

This patient was one of the first to benefit from vinblastine at this institution and he had a remarkable response. In all, the drug provided benefit throughout a 30 month period and he was in complete remission much of the time. His only

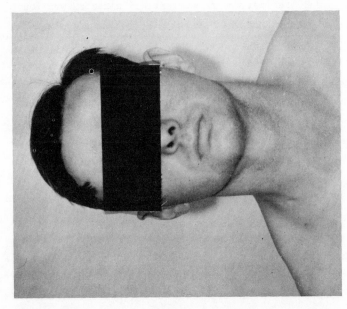

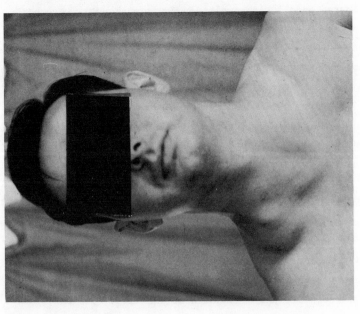

A.

B.

Case 1

Fig. 3

Marked reduction in supraclavicular adenopathy one month
after initiation of vinblastine therapy for generalized Hodgkin's Disease.

significant drug-related complication was neuritis which cleared with cessation of therapy and did not reappear when drug was begun again. He also demonstrated the need for determining the frequency of injections for each patient; he began to relapse when drug was administered every 3 weeks but was responsive to injections given every 2 weeks.

Case 2

This 45 year old black male presented to the Yale-New Haven Medical Center on July 31, 1961 with a 3-4 month history of painless lymph node enlargement in the neck, axilla, and groin. Lymph node biopsy revealed lymphocytic lymphoma and bone marrow aspiration showed heavy infiltration by lymphocytes. He was considered to have leukolymphosarcoma and was treated with 0.3 mg/kg of mechlorethamine HCl with very little diminution in the lymphadenopathy. The peripheral white blood cell count of $28,000/mm^3$ with 89% lymphocytes fell to a nadir of $6800/mm^3$ with 45% lymphocytes as a result of therapy during the ensuing 2-3 weeks after treatment.

On August 4, 1961 he was begun on weekly injections of vinblastine sulfate in dosages beginning at 0.1 mg/kg and gradually increasing to 0.3 mg/kg in the next 2 months. Treatment was associated with minimal nausea and some pain in his cervical nodes; tumor regression was definite (Figure 4). To attempt to improve therapeutic effectiveness, dosage was gradually increased to 0.7 mg/kg without ill effects and only gradual anemia. He gained weight and was able to return to work; maximal improvement was seen in January 1962, and to avoid further anemia and leukopenia, drug was discontinued in March. He did well until June when his adenopathy increased; vinblastine was begun again and once more tumor regression was induced. Again he experienced temporary pain in the lesions. The white blood cell count was maintained in the normal range and differential counts revealed

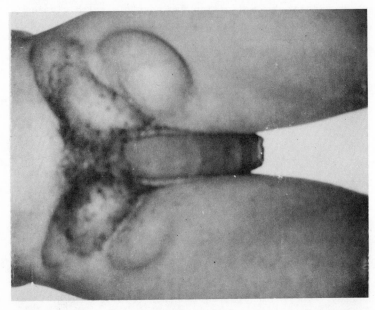

Case 2

Fig. 4

A. August 1961 before therapy with vinblastine, massive inguinal adenopathy with penile edema.

A.

B.

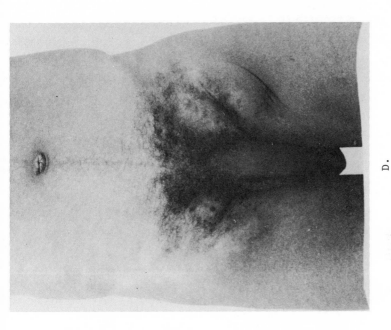

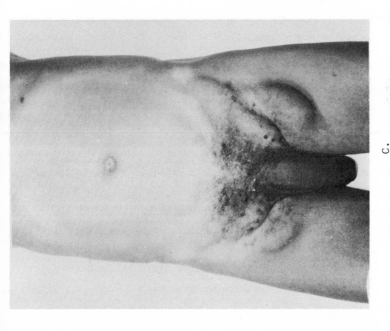

D.

C.

Case 2

Fig. 4

C.   December 1961, four months after vinblastine therapy showing further regression of adenopathy.

D.   January 1962, relatively stable lymphadenopathy after vinblastine. Lesions remained much like this except for temporary relapse until June 1963.

45-60% polymorphonuclear leukocytes. He was treated on a 2-week schedule with good results until he began to develop weakness. This was attributed to drug toxicity and the dosage was reduced from 0.5 to 0.25 mg/kg with improvement in muscle strength. This program was continued successfully until June 1963 when nodal enlargement and hepatosplenomegaly could not be controlled. He expired in March 1964.

Comment:

This case presents a number of unusual aspects. This patient represented an extraordinarily good response to vinblastine in leukolymphosarcoma both in terms of degree of response and its long duration (20 months). He also tolerated a very high dosage of vinblastine without marked neurologic effects and only developed muscle weakness after receiving a great amount of drug. He clearly demonstrated on two occasions pain in the tumor site during lysis by drug. In general, one would not now give such high dosages of vinblastine but would continue therapy in the 0.1-0.35 mg/kg range. Furthermore in 1971 the demonstrated superior responses to vincristine makes this agent a more rational choice than vinblastine, particularly in a program of combination chemotherapy.

Case 3

This 27 year old white male construction worker entered the Yale-New Haven Hospital on December 13, 1961 with complaints of intermittent fever, weakness and edema of the leg. Two years before, a cervical node biopsy had revealed Hodgkin's Disease; he had been considered to have generalized disease and received mechlorethamine HCl and chlorambucil, with minimal benefit, in May 1961. Physical examination revealed diffuse lymphadenopathy in the neck, axillae and groin as well as marked ascites. Vinblastine was begun at a dose of 0.4 mg/kg with the subsequent

development of marked leukopenia ($800/mm^3$) a week later. He had
a partial response in nodal disease to this first dose and 2
weeks later (after hematopoietic recovery) he was continued as
an outpatient on weekly doses of 0.2 mg/kg without myelosup-
pression. As his ascites cleared, hepatosplenomegaly became
evident and gradually regressed with continued therapy. Within
a month he returned to work, and after another month virtually
all his adenopathy had disappeared. Vinblastine was continued
every 2 weeks with no difficulty except for occasional mild
leukopenia until October 1962 when nodes reappeared. Weekly
administration of drugs was unsuccessful, his relapse worsened,
and in December, vinblastine was discontinued (total remission
duration of vinblastine about 10 months). He was admitted to
the hospital once more with generalized lymphadenopathy, chylous
ascites, pleural effusion and peripheral edema. Vincristine
was given weekly in dosages of 0.03-0.05 mg/kg with some diminution in ascites, but, because of the slowness of the response,
X-ray therapy to the abdomen and mediastinum was substituted.
This appeared to achieve little and vincristine was continued.
Chest X-ray continued to reveal opacification of the left lung
by fluid (See Figure 5A). For the next 14 weeks, drug was
given at a usual dose of 0.05 mg/kg weekly. On this program
his girth and weight decreased as fluid accumulation lessened.
His condition and performance states improved steadily (See
Figure 6). He gradually became aware of paresthesias in his
finger tips, the deep tendon reflexes disappeared and he developed double vision. The chest X-ray (Figure 5B) demonstrated
absence of fluid and in view of his good clinical condition
and signs of drug toxicity, dosage was reduced to 0.03 mg/kg
at 2 week intervals. His paresthesias improved during the
next 2 months and the double vision disappeared. He steadily
gained weight and did well except for several bouts of constipation which were relieved by laxatives. He continued in

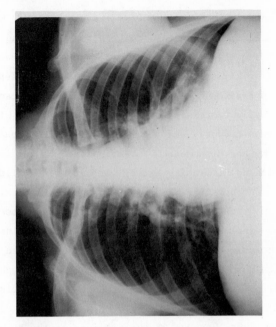

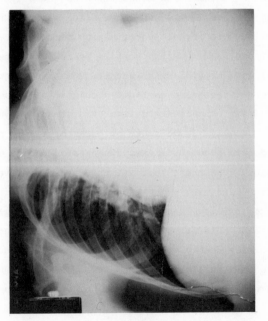

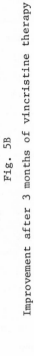

Case 3

Fig. 5A

Chylous pleural effusion

Fig. 5B

Improvement after 3 months of vincristine therapy

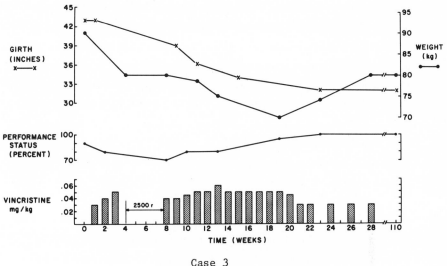

Case 3

Fig. 6

Clinical manifestations of resolution of ascites reflected in improvements in performance status, girth and weight after vincristine therapy.

complete remission with minimal discomfort from the intermittent tingling of his finger tips until March 1965 when he again developed hepatosplenomegaly, constitutional symptoms and weight loss. He derived only short-term benefits from subsequent therapy with cyclophosphamide, prednisone and procarbazine and expired in December 1965.

Comment:

This patient sustained benefit from vinblastine therapy for 10 months after he had become resistant to alkylating agents. Advanced generalized disease was controlled with only transient leukopenia once his dosage of drug was titrated. Occasionally increasing the frequency of drug administration will reverse impending relapse but this did not benefit this patient.

Despite his resistance to vinblastine, he responded steadily and very well to vincristine administration, demonstrating a lack of clinical cross-resistance in these closely related drugs. His response to vincristine was striking for its completeness and long duration (about 2 years). Neurologic toxicity was kept within acceptable limits by modification of drug dosage and schedule. The benefits this man enjoyed as a result of chemotherapy were largely achieved by these two alkaloids; in contrast, alkylating agents were of only short-term benefit.

Case 4

This 26 year old man with acute lymphocytic leukemia was treated with increasing weekly doses of vincristine after he had become resistant to the antineoplastic effects of methotrexate, 6 mercaptopurine and steroids. Only as the dosage of vincristine was raised from 0.04 to 0.07 mg/kg was the marked thrombocytopenia and splenomegaly influenced by drug and he experienced a sudden and dramatic fall in the peripheral white blood count and disappearance of lymphoblasts from the circulation. (See Figure 7) Coincident with this improvement, he developed renal colic with uric acid nephropathy that required vigorous efforts at hydration to successfully prevent compromise of renal function. He was maintained for several weeks on 0.04 mg/kg vincristine when he developed a unilateral oculomotor nerve palsy. Vincristine neurotoxicity was suspected but lumbar puncture revealed meningeal leukemia; shortly thereafter he relapsed with return of splenomegaly and blasts in the peripheral blood.

Comment:

Since this patient was treated, ample data have accumulated confirming the benefits of combination drug therapy for patients with leukemia and few should be treated with vincristine alone.

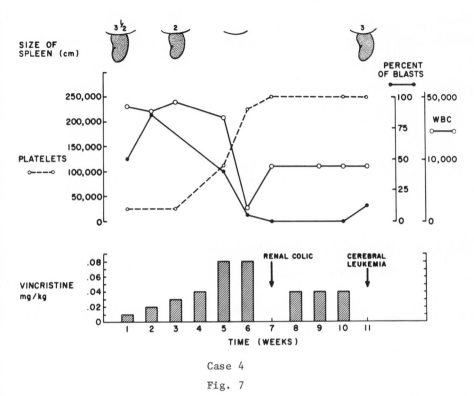

Case 4

Fig. 7

Dramatic but brief response of acute lymphocytic leukemia to vincristine.

While vincristine is a very good agent for inducing remission, particularly in acute lymphocytic leukemia, it has little value for maintenance of remission and other drugs are used for this purpose. This case demonstrates three other points: first, the usual necessity for high dosages of drug to achieve response in this disease; second, the availability of rapidly effective chemotherapy creates dangers of a sudden increase in urate load for the kidney and such patients should be hydrated and treated with allopurinol to avoid this complication;[115] third, that symptoms suggestive of drug toxicity should be com-

pletely investigated before they are accepted as such.   In
leukemia, nerve palsy more often suggests meningeal involvement
than side effects of vincristine.   Similarly paresthesias may
suggest either extradural cord compression or drug toxicity
and a careful clinical analysis may be necessary to determine
which is responsible.

REFERENCES

1. I.S. Johnson, J.G. Armstrong, M. Gorman, and J.P. Burnett,
    Jr., Cancer Res., 23, 1390 (1963)

2. W.H. Bond, R.J. Rohn, L.H. Bates and M.E. Hodes, Cancer,
    19, 213 (Feb. 1966).

3. M.E. Hodes, R.J. Rohn, W.H. Bond, and J. Yardley, Cancer
    Chemother. Rep., 28, 53 (1963).

4. G. Mathe, M. Schneider, P. Band, J.L. Amiel, L. Schwarzen-
    berg, A. Cattan, and J.R. Schlumberger, Cancer Chemother.
    Rep., 49, 47 (1965).

5. S.D. Gailani, J.G. Armstrong, P.P. Carbone, C. Tan, and
    J.F. Holland, Cancer Chemother. Rep., 50, 95 (1966).

6. J.G. Armstrong, R.W. Dyke, P.J. Fouts, J.J. Hawthorne,
    C.J. Jansen, Jr., and A.M. Peabody, Cancer Res., 27, 221
    (1967).

7. I.S. Johnson, Cancer Chemother. Rep., 52, 455 (1968).

8. C.G. Smith, J.E. Grady, and F.P. Kupiecki, Cancer Res.,
    25, 241 (1965).

9. F.A. Valeriote and W.R. Bruce, J. Nat. Cancer Inst., 35,
    851 (1965).

10. L. Morasca, C. Rainisio, and G. Masera, Eur. J. Cancer, 5,
    79 (1969).

11. W.A. Creasey, Cancer Chemother. Rep., 52, 501 (1968).

12. E.K. Wanger and B. Roizman, Science, 162, 569 (1968).

13. M.J. Cline, Brit. J. Haematol., 14, 21 (1968).

14. M.J. Cline, Blood, 30, 176 (1967).

15. M.J. Cline and E. Rosenbaum, Cancer Res., 28, 2516 (1968).

16. H.C. Laurie and M.L.N. Willoughby, Brit. J. Haematol., 17, 251 (1969).

17. D. Roberts, T.C. Hall, and D. Rosenthal, Cancer Res., 29, 789 (1969).

18. M.E. Hodes, R.J. Rohn, W.H. Bond, J.M. Yardley, and W.S. Corpening, Cancer Chemother. Rep., 16, 401 (1962).

19. J.G. Armstrong, R.W. Dyke, P.J. Fouts, and J.E. Gahimer, Cancer Chemother. Rep., 18, 49 (1962).

20. T.L. Wright, J. Hurley, D.R. Korst, R.W. Monto, R.J. Rohn, J.J. Will and J. Louis, Cancer Res., 23, 169 (1963).

21. J.H. Robertson, and G.M. McCarthy, Lancet, 2, 353 (1969).

22. M.S. Rose, Lancet, 1, 213 (1967).

23. B.C. Lampkin, Lancet, 1, 891 (1969).

24. P.P. Carbone, V. Bono, E. Frei, III and C.O. Brindley, Blood, 21, 640 (May 1963).

25. E. Lefkowitz, R.C. DeConti, P. Calabresi, (unpublished data).

26. W. Tobin, S.G. Sandler, Cancer Chemother. Rep., 52, 519 (1968).

27. S.G. Sandler, W. Tobin, E.S. Henderson, Neurology, 19, 367 (1969).

28. P.G. Gottschalk, P.J. Dyck, J.M. Kiely, Neurology, 18, 875 (1968).

29. R.A. Bohannon, D.G. Miller, and H.D. Diamond, Cancer Res., 23, 613 (1963).

30. R.K. Shaw and J.A. Bruner, Cancer Chemother. Rep., 42, 45 (1964).

31. R.N. Fine, R.R. Clarke and N.A. Shore, Am. J. Dis. Child, 112, 256 (1966).

32. L.M. Slater, R.A. Wainer and A.A. Serpick, Cancer, 23, 122 (1969).

33. G. Owens, R. David, L. Belmusto, M. Bender and M. Blau, Cancer, 18, 756 (1965).

34. J.J. Weitzman, C.A. Brubaker, N. Hastings and W.H. Snyder, Jr., J. Pediat. Surg., 1, 368 (1966).

35. J.V. Apostol, P.F. Nora and F.W. Preston, Cancer Chemother. Rep., 50, 573 (1966).

36. S.S. Schochet, Jr., P.W. Lampert and K.M. Earle, J. Neuropathol. Exp. Neurol., 27, 645 (1968).

37. T. Makinodan, G.W. Santos and R.P. Quinn, Pharmacol. Rev., 22, 189 (1970).

38. C. Radzikowski, J.P. Glynn and A. Goldin, Cancer Chemother. Rep., 32, 15 (1963).

39. A.C. Aisenberg and B. Wilkes, J. Clin. Invest., 43, 2394 (1964).

40. J.E. Harris, R. Alexanian, E.M. Hersh, W.M. Leary, Can.Med. Ass. J., 101, 231 (1969).

41. E.M. Hersh, P.P. Carbone, V.G. Wong and E.J. Freireich, Cancer Res., 25, 997 (1965).

42. B. Belonorsky, J. Siracky, L. Sandor and E. Klauber, Neoplasma, 7, 397 (1960).

43. R. Richter, J.C. Calamera, M.C. Morgenfeld, A.L. Kierszenbaum, J.C. Lavieri and R.E. Mancini, Cancer, 25, 1026 (1970).

44. L.G. Sobrinho, R.A. Levine, and R.C. DeConti, Amer. J. Obstet. and Gynaec. 108, 135 (1971).

45. M.S. Joshi and R.Y. Ambaye, Indian J. Exp. Biol. 6, 256 (1968).

46. V.H. Fern, Science, 141, 426 (1963).

47. S.Q. Cohlan and D. Kitay, J. Pediat., 66, 541 (1965).

48. J.G. Armstrong, R.W. Dyke, P.J. Fouts and C.J. Jansen, Ann. Intern. Med., 61, 106 (1964).

49. A.I. Rosenzweig, Q.E. Crews, Jr., and H.G. Hopwood, Ann. Intern. Med., 61, 108 (1964).

50. M.J. Lacher, Ann. Intern. Med., 61, 113 (1964).

51. H.O. Nicholson, J. Obstet. Gynaec. Brit. Emp., 75, 307 (1968).

52. E.Z. Ezdinli and L. Stutzman, Cancer, 22, 473 (1968).

53. J.L. Scott, Cancer Chemother. Rep., 27, 27 (1963).

54. W.D. Sohier, Jr., R.K.L. Wong and A.C. Aisenberg, Cancer, 22, 467 (1968).

55. A.M. Jelliffe, Brit. J. Cancer, 23, 44 (1969).

56. L. Stutzman, E.Z. Ezdinli and M.A. Stutzman, J. Amer. Med. Ass., 195, 173 (1966).

57. C.L. Spurr and P.P. Carbone, Cancer Res., 28, 811 (1968).

58. F.A. Flatow, J.E. Ultmann, G.A. Hyman and F.M. Muggia, Cancer Thermother. Rep., 53, 39 (1969).

59. B.S. Morse and F. Stohlman, Jr., J. Clin. Invest, 45, 241 (1966).

60. D. Burkitt, Cancer, 19, 1131 (1966).

61. H.M. Williams, Proc. Amer. Ass. Amer. Res, 5, 68 (1964). (abstract)

62. M.J. Lacher and J.R. Durant, Ann. Intern. Med., 62, 468 (1965).

63. V.J. DeVita, Jr., A.A. Serpick and P.P. Carbone, Ann. Intern. Med., 73, 881 (1970).

64. A. Repoport, P. Cole and J. Mason, Cancer, 24, 377 (1969).

65. M. Levitt, R.C. DeConti, H.A. Pearson, J.C. Marsh, R.P. Zanes, Jr., M.S. Mitchell, H.W. Kaetz and J.R. Bertino, Conn. Med., 34, 862 (1970).

66. L.R. Farber, M. Levitt, R.C. DeConti, M.S. Mitchell, J.C. Marsh, R.T. Skeel, H.A. Pearson, R.P. Zanes and J.R. Bertino, Proc. Amer. Ass. Cancer Res., 12, 87 (1971).

67. J.H. Moxley, III, V.T. DeVita, K. Brace and E. Frei, III, Cancer Res., 27, 1258 (1967).

68. S. Lowenbraun, V.T. DeVita and A.A. Serpick, Cancer, 25, 1018 (1970).

69. B. Hoogstraten, A.H. Owens, R.E. Lennard, O.J. Glidewell, L.A. Leone, K.B. Olson, J.B. Harley, S.R. Townsend, S.P. Miller and C.L. Spurr, Blood, 33, 370 (1969).

70. M. Levitt, R.C. DeConti, J.C. Marsh, M.S. Mitchell, R.T. Skeel, L.R. Farber and J.R. Bertino, (Submitted to Amer. Soc. of Clin. Oncology Meetings in Chicago, 1971).

71. J.F. Holland, Cancer Res., 29, 2270 (1969).

72. E.S. Henderson and R.J. Samaha, Cancer Res., 29, 2272 (1969).

73. W.W. Satow, T.J. Vietti, D.J. Fernbach, D.M. Lane, M.H. Donaldson and Daisilee H. Berry, J. Pediat., 73, 426 (1968).

74. J. Bernard, M. Boiron, C. Jacquillat and M. Weil, (Abstract of the Simultaneous Sessions XII Congress. Intern. Soc. Hemat.) 5, 5 (1968).

75. G. Mathe, M. Hayat, and L. Schwarzenberg, Lancet, 2, 380 (1967).

76. E.J. Freireich, E.S. Henderson, M.R. Karon and E. Frei, III, in, *The Proliferation and Spread of Neoplastic Cells*, The Williams & Wilkins Co., Baltimore (1968).

77. J.P. Howard, Cancer Chemother. Rep., 51, 465 (1967).

78. D.M. Lane, M.E. Haggard, D. Lonsdale, K. Starling and M.P. Sullivan, Cancer Chemother. Rep., 54, 113 (1970).

79. K. Starling, D.M. Lane, W.W. Sutow, R.W. Monto and W.G. Thurman, Cancer Chemother. Rep., 54, 292 (1970).

80. G.P. Canellos and J. Whang-Peng, Clin. Res., 17, 400 (1969).

81. P.P. Carbone, J.H. Tjio, J. Whang, J.B. Block, W.B. Kremer and E. Frei, III, Ann. Intern. Med., 59, 622 (1963).

82. W.W. Sutow, Cancer Chemother. Rep., 52, 485 (1968).

83. O.S. Selawry, J.F. Holland and I.J. Wolman, Cancer Chemother. Rep., 52, 497 (1968).

84. J.A. Wolff, W. Krivit, W.A. Newton, Jr. and G.J. D'Angio, N. Engl. J. Med., 279, 290 (1968).

85. M.P. Sullivan, Cancer Chemother. Rep., 52, 481 (1968).

86. W.W. Sutow, Cancer, 18, 1585 (1965).

87. S.B. Kontras and W.A. Newton, Jr., Cancer Chemother. Rep.,
     12, 39 (1961)

88. D. Pinkel, Cancer, 15, 42 (1962).

89. W.G. Thurman, D.J. Fernbach and M.P. Sullivan, The Writing
     Comm. of the Pediat. Div. of the S.W. Ca. Chemother. Stdy.
     Group, N. Engl. J. Med., 270, 1336 (1964).

90. J. Windmiller, D.H. Berry, T.B. Haddy, T.J. Vietti and
     W.W. Sutow, Amer. J. Dis. Child., 111, 75 (1966).

91. D.H. James, Jr., O. Hustu, E.L. Wrenn, Jr., and D. Pinkel,
     J. Amer. Med. Ass., 194, 123 (1965).

92. C.B. Pratt, D.H. James, Jr., C.P. Holton and D. Pinkel,
     Cancer Chemother. Rep., 52, 489 (1968).

93. A.E. Evans, R.M. Heyn, W.A. Newton, Jr., and S.L. Leikin,
     J. Amer. Med. Ass., 207, 1326 (1969).

94. A. Sawitsky and F. Desposito for Acute Leukemia, Group B,
     Cancer Chemother. Rep., 53, 93 (1969).

95. N.J. Gubisch, D. Norena, C.P. Perlia and S.G. Taylor, III,
     Cancer Chemother. Rep., 32, 19 (1963).

96. R.J. Reitemeier, C.G. Moertel and C.M. Blackburn, Cancer
     Chemother. Rep., 34, 21 (1964).

97. R. Hertz, M.B. Lipsett and R.H. Moy, Cancer Res, 20, 1050
     (1960).

98. C.R. Smart, D.B. Rochlin, A.M. Nahum, A. Silva and D.
     Wagner, Cancer Chemother. Rep., 34, 31 (1964).

99. B. Johnston and E.T. Novales, Philipp. J. Cancer, 4, 41
     (1962).

100. H.M. Williams, Proc. Amer. Ass. Cancer Res., 6, 68 (1964).

101. C.R. Smart, R.E. Ottoman, D.B. Rochlin, J. Hornes, A.R.
     Silva and H. Goepfert, Cancer Chemother. Rep., 52, 733
     (1968).

102. A. Mittelman, R. Grinberg and T. Dao, Cancer Chemother. Rep., 34, 25 (1964).

103. C.W. MacFarlane, B.J. Doughty and W.A. Crosbie, Brit. J. Dis. Chest, 56, 64 (1962).

104. A.B. Myles, Brit. J. Cancer, 20, 264 (1966).

105. D.H. James, Jr., O. Hustin, E.L. Wrenn, Jr., and D. Pinkel, Proc. Am. Ass. Cancer Res., 7, 34 (1966).

106. R.H. Chanes and R.H. Bottomley, Proc. Am. Ass. Cancer Res., 9, 12 (1968).

107. L. Nathanson, T.C. Hall, A. Schilling and S. Miller, Cancer Res., 29, 419 (1969).

108. J. Horton, K.B. Olson, P. Gehart and M. Spear, Cancer Chemother. Rep., 49, 59 (1965).

109. D.E. Kayhoe, Minutes of the Eastern Cooperative Group for Solid Tumor Chemotherapy, Nov. 13, 1964.

110. C. Nervi, C. Casale and M. Cortese, Tumori, 55, 103 (1969).

111. J. Solomon, J.L. Steinfeld and J.R. Bateman, Cancer , 20 747 (1967).

112. D. Mendelson, Cancer Chemother. Rep., 53, 90 (1969).

113. J.H. Moon, Cancer Chemother. Rep., 53, 91 (1969).

114. R.G. Cooper, Proc. Am. Ass. Cancer Res., 10, 15 (1969).

115. R.C. DeConti and P. Calabresi, N. Engl. J. Med., 274, 481 (1966).

AUTHOR INDEX

Numbers in parentheses are reference numbers and indicate that
an author's work is referred to although his or her name is not
cited in the text. Underlined numbers give the page on which
the complete reference is listed.

A

Abraham, D.J., 69(65), 83, 86
  (5), 88(16), 90(23), 91(16,
  31, 32, 33, 34, 35), 92(16,
  37, 38, 39, 40), 103(5),
  106(5), 107(5), 110(5), 111
  (5), 116, 117, 118, 131,
  132(22,25), 133(21), 140
Adamson, R.H., 220(47), 233
Afzelius, B., 226(86), 235
Agrifoglio, M.F., 210(8), 223
  (8), 232
Aguilar-Santos, G., 114(123,
  124), 122
Agustin, B.M., 215, 216(31),
  233
Ahmad, A., 142(13), 144(13),184
Ahond, A., 63(79), 83
Aisenberg, A.C., 245(39), 247
  (54), 259(54), 274, 275
Albert, O., 71, 83
Aldaba, L., 86(9), 116
Alexanian, R., 245(40), 274
Allen, F.H., 152(31), 157(31),
  185
Allen, J., 214(23), 216(23),
  232
Altman, R.F.A., 107(109), 121
Ambaye, R.Y., 246(45), 274
Amiel, J.L., 238(4), 272
Amoure, J.E., 220(46), 233
Anon, 88(14), 116
Aoki, Y., 164(60), 187
Apostol, J.V., 245(35), 274
Arai, T., 153(36), 185

Arendell, J.P., 214(26), 216
  (26), 232
Arigoni, D., 144(15), 145(21),
  147(25), 154(10), 155(43,
  45), 160(21,55), 162(55,56,
  57), 168(56), 180(56), 181
  (56), 184, 186
Armstrong, J.G., 69(67), 83,
  125(9), 139, 209(6), 210
  (13), 220(13), 231, 232,
  238(1,5,6), 242(19), 246
  (48), 272, 273, 274
Arndt, R.R., 106(64), 119
Arreguin, B., 195(11), 207
Arthur, H.R., 106(76), 107(76),
  120
Aynilian, G.H., 94(41,42,43),
  100(41), 101(41,42,43), 114
  (41,42,43), 118

B

Babcock, P.A., 199, 200, 202
  (45), 204, 208
Bader, R.F., 107(101), 121
Baker, L.A., 125(7), 139
Band, P., 238(4), 272
Bariety, M., 221, 234
Barnes, A.J., Jr., 60(29, 32,
  33, 49), 62(29,32), 63(29),
  64(32, 33, 49), 65(32, 49),
  81, 103(53, 58), 118, 119
Baronowsky, P., 220(51), 234
Basu, N.K., 59(25), 80
Bateman, J.R., 260(111), 278
Bates, L.H., 238(2), 272
Batlori, L., 106(83), 120

279

SUBJECT INDEX

A

N-Acetyl-vincoside,
   isolation from *Catharanthus
      roseus* seedlings, 66,
      159
Actinomycin D,
   in combination chemotherapy
      of rhabdomyosarcoma, 258
   for Wilms' tumor, 257
2-Acylindole alkaloids,
   reactions with ceric ammonium
      sulfate, 93
Administration routes, human,
   for leurocristine, 242
   for vincaleukoblastine, 241,
      242
Adrenergic blocking effects of
   leurosine, 72
Agrobacterium tumefaciens as an
   inducer of crown gall in
   Catharanthus roseus, 193
Ajmalicine,
   antidiuretic effects of, 89
   antiviral effects of, 90
   biosynthesis of, 147, 148,
      169
   biosynthetic conversion to
      geissoschizine,
   diuretic effects of, 74
   hypoglycemic effects of, 89
   incorporation of geraniol
      derivatives into, 161
   isolation from,
      Catharanthus lanceus, 91,
         98, 104, 110
      Catharanthus longifolius,
         124
      Catharanthus pusillus, 104,
         106, 108, 110
      Catharanthus roseus, 54, 59,

66, 106, 110
   Catharanthus trichophyllus,
      104, 106, 111, 113, 124
presence in,
   Catharanthus roseus seed-
      lings, 66
   Mitragyna javanica, 106
   Mitragyna javanica var.
      microphylla, 106
   Mitragyna speciosa, 106
   Pausinystalia yohimbe, 106
   Rauvolfia caffra, 106
   Rauvolfia canescens, 106
   Rauvolfia chinensis, 106
   Rauvolfia heterophylla,
      106
   Rauvolfia micrantha, 106
   Rauvolfia sellowii, 106
   Rauvolfia serpentina, 106
   Rauvolfia tetraphylla, 106
   Rauvolfia verticillata,
      106
   Rauvolfia vomitoria, 106
   Rauvolfia yunanensis, 106
   Stemmadenia obovata, 106
loganin incorporation into,
   151, 166
mevalonic acid incorporation
   into, 148
muscle relaxant effects of,
   72
physical data for, 104
precursor for,
   catharanthine, 179
   serpentine, 179
   vindoline, 179
presence in *Catharanthus
   roseus* tissue cultures,
   205
secologanin incorporation
   into, 151

293

solid tumors, 249
testicular tumors, 260
Wilms' tumor, 257
side effects in humans by
   therapy with leurocristine or vincaleukoblastine include,
abdominal pain, 244
alopecia, 244
constipation, 244
cranial nerve palsy, 244
depression, 244
generalized weakness, 244
hoarseness, 244
hyponatremia, 244
jaw pain, 244
leg cramps, 244
paresthesia, 243, 244
perforated colon, 244
peripheral neuropathy, 242
phlebitis, 244
slapping gait, 244
side effects on i.v. use,
   241, 243, 244
side effects, lack of, 244
sites of action for, 230
spermatogenesis altered by,
   246
structure of, 67
structure determination of,
   130
teratogenic effects of, 246
therapy for,
   acute leukemias, 249, 254,
      255, 256
   CNS tumors, 257
   Ewing's sarcoma, 257
   histiocytosis X, 257
   Hodgkin's disease, 248, 249
   lymphosarcoma, 249, 250
   malignant teratoma, 257
   multiple myeloma, 249
   neuroblastoma, 257
   reticulum cell sarcoma,
      249, 251
   retinoblastoma, 257
   rhabdomyosarcoma, 257
   soft tissue sarcomas, 257
   solid tumors, 249

Wilms' tumor, 257
yields from *Catharanthus
   roseus*, 56, 57
X-ray structure determination of, 129
Leurosidine,
   antileukemic activity of, 58
   in cancer therapy, 238
   CNS depressant effects of,
      72
   CNS stimulant effects of, 72
   effects on nucleic acid
      metabolism, 214, 216
   isolation from *Catharanthus
      roseus*, 57, 64
   side effects in humans, 238
   structure of, 68
   structure elucidation of,
      130
Leurosine,
   adrenergic blocking effects
      of, 72
   antileukemic activity of, 58
   antiviral activity of, 90
   biochemical effects of, 214
   clinical use of,
      methiodide, 238
      sulfate salt, 238
   cytotoxicity of, 88
   effects on nucleic acid meta-
      bolism, 214, 216
   effects on lipid metabolism,
      220, 221
   hypoglycemic activity of, 71,
      77, 89
   isolation from,
      Catharanthus lanceus, 91,
         98, 110
      Catharanthus longifolius,
         124
      Catharanthus ovalis, 124
      Catharanthus pusillus, 103,
         110
      Catharanthus roseus, 53,
         54, 57, 64, 103, 110
   mass spectrum of, 132
   NMR spectrum of, 136
   physical data summary for,
      105